DEVELOPING ASSESSMENTS FOR THE NEXT GENERATION SCIENCE STANDARDS

Committee on Developing Assessments of Science Proficiency in K-12

James W. Pellegrino, Mark R. Wilson, Judith A. Koenig, and Alexandra S. Beatty, *Editors*

Board on Testing and Assessment

Board on Science Education

Division of Behavioral and Social Sciences and Education

NATIONAL RESEARCH COUNCIL
OF THE NATIONAL ACADEMIES

THE NATIONAL ACADEMIES PRESS
Washington, D.C.
www.nap.edu

THE NATIONAL ACADEMIES PRESS   500 Fifth Street, NW   Washington, DC   20001

NOTICE: The project that is the subject of this report was approved by the Governing Board of the National Research Council, whose members are drawn from the councils of the National Academy of Sciences, the National Academy of Engineering, and the Institute of Medicine. The members of the committee responsible for the report were chosen for their special competences and with regard for appropriate balance.

This study was supported by the S. D. Bechtel, Jr. Foundation, the Carnegie Corporation of New York under Contract No. B8834, and the William and Flora Hewlett Foundation under Contract No. 2012-7436. Any opinions, findings, conclusions, or recommendations expressed in this publication are those of the authors and do not necessarily reflect the views of the organizations or agencies that provided support for the project.

International Standard Book Number-13:   978-0-309-28951-1
International Standard Book Number-10:   0-309-28951-3
Library of Congress Control Number: 2014937574

Additional copies of this report are available from the National Academies Press, 500 Fifth Street, NW, Keck 360, Washington, DC 20001; (800) 624-6242 or (202) 334-3313; http://www.nap.edu.

Printed in the United States of America

Suggested citation: National Research Council. (2014). *Developing Assessments for the Next Generation Science Standards*. Committee on Developing Assessments of Science Proficiency in K-12. Board on Testing and Assessment and Board on Science Education, J.W. Pellegrino, M.R. Wilson, J.A. Koenig, and A.S. Beatty, *Editors*. Division of Behavioral and Social Sciences and Education. Washington, DC: The National Academies Press.

THE NATIONAL ACADEMIES
Advisers to the Nation on Science, Engineering, and Medicine

The **National Academy of Sciences** is a private, nonprofit, self-perpetuating society of distinguished scholars engaged in scientific and engineering research, dedicated to the furtherance of science and technology and to their use for the general welfare. Upon the authority of the charter granted to it by the Congress in 1863, the Academy has a mandate that requires it to advise the federal government on scientific and technical matters. Dr. Ralph J. Cicerone is president of the National Academy of Sciences.

The **National Academy of Engineering** was established in 1964, under the charter of the National Academy of Sciences, as a parallel organization of outstanding engineers. It is autonomous in its administration and in the selection of its members, sharing with the National Academy of Sciences the responsibility for advising the federal government. The National Academy of Engineering also sponsors engineering programs aimed at meeting national needs, encourages education and research, and recognizes the superior achievements of engineers. Dr. C. D. Mote, Jr., is president of the National Academy of Engineering.

The **Institute of Medicine** was established in 1970 by the National Academy of Sciences to secure the services of eminent members of appropriate professions in the examination of policy matters pertaining to the health of the public. The Institute acts under the responsibility given to the National Academy of Sciences by its congressional charter to be an adviser to the federal government and, upon its own initiative, to identify issues of medical care, research, and education. Dr. Harvey V. Fineberg is president of the Institute of Medicine.

The **National Research Council** was organized by the National Academy of Sciences in 1916 to associate the broad community of science and technology with the Academy's purposes of furthering knowledge and advising the federal government. Functioning in accordance with general policies determined by the Academy, the Council has become the principal operating agency of both the National Academy of Sciences and the National Academy of Engineering in providing services to the government, the public, and the scientific and engineering communities. The Council is administered jointly by both Academies and the Institute of Medicine. Dr. Ralph J. Cicerone and Dr. C. D. Mote, Jr., are chair and vice chair, respectively, of the National Research Council.

www.national-academies.org

COMMITTEE ON DEVELOPING ASSESSMENTS OF SCIENCE PROFICIENCY IN K-12

Preface

The U.S. public education system has seen many reform efforts come and go, and the claim that few leave lasting benefits after they are out of fashion can be made with some credibility. The 2012 *A Framework for K-12 Science Education: Practices, Crosscutting Concepts, and Core Ideas* offers the promise of something quite different. The framework proposed a dramatic rethinking of science education grounded in a thoughtful analysis of the reasons science education has fallen short. With its insistence that science education integrate the practices, disciplinary core ideas, and crosscutting concepts of science and engineering in a coherent fashion across the K-12 years, the framework established goals that cannot be achieved through tinkering. Implementing its vision will require a thorough rethinking of each element of science education, including science assessment.

Assessments, understood as tools for tracking what and how well students have learned, play a critical role in the education system—from classrooms to state houses. Frequent misapplication of these tools and misuse of their results have tarnished their reputation. But the new K-12 framework makes clear that such tools, reflecting new modes of assessment designed to measure the integrated learning it envisions, will be essential. Our committee was asked to develop an approach to science assessment that would support attainment of this vision as it has been elaborated in the *Next Generation Science Standards: For States, By States*, which were developed in response to the framework. Both documents are new, and the changes they call for are barely under way, but new assessments will be needed as soon as states and districts begin the process of implementing the

Next Generation Science Standards (NGSS) and changing their approach to science education. This meant that our committee had to work quickly to assemble and evaluate a wide range of information related to research and practice and to assimilate thinking and perspectives from across many disciplines.

With funding from the S. D. Bechtel, Jr. Foundation, the Carnegie Corporation of New York, and the William and Flora Hewlett Foundation, the National Research Council (NRC) established the Committee on Developing Assessment of Science Proficiency in K-12 to carry out a consensus study under the aegis of the Board on Testing and Assessment and the Board on Science Education. The committee was asked to recommend strategies for developing assessments that measure student proficiency in science as laid out in the new K-12 science education framework.

The committee benefited from the work of many others, and we wish to thank the many individuals who assisted us. We first thank the sponsors who supported this work: the S. D. Bechtel, Jr. Foundation, the Carnegie Corporation of New York, and the William and Flora Hewlett Foundation. We particularly thank the representatives from the sponsoring groups for their ongoing assistance and insights about the project: Andres Henriquez with Carnegie; Dennis Udall with Hewlett; and Soo Venkateson with Bechtel.

During the course of its work, the committee met four times, including two public sessions. The first public session was held in Palo Alto, California, at the offices of the Moore Foundation. We thank the staff at the Moore Foundation, particularly Janet Coffey, for their gracious hospitality in hosting this meeting. At this meeting, we heard from representatives of the two Race to the Top Assessment Program consortia with regard to their plans for using computer-based assessments, performance tasks, and other innovative approaches to assessing English language arts and mathematics that might be applied to assessment of science. We thank Jeff Nelhaus and Enis Dogan for their presentations on the work of the Partnership for Assessment of Readiness for College and Careers, and we thank Joe Wilhoft and Stan Rabinowitz for their presentations on the work of the Smarter Balanced Assessment Consortia.

The second meeting included a public workshop designed for the committee to learn more about innovative approaches to science assessment. We thank Alan Friedman, former member of the National Assessment Governing Board, and Peggy Carr, with the National Center for Education Statistics, for their presentation about the Computer Interactive and Hands-On Science Assessment of the National Assessment of Educational Progress; Rosemary Reshetar, with the

College Board, for her presentation about the newly revised Advanced Placement assessment in biology; Edys Quellmalz, with WestEd, for her presentation about the SimScientist Program; and Joseph Krajcik, with Michigan State University, for his presentation about the Investigating and Questioning our World through Science and Technology Program.

The workshop also provided time for the committee to learn more about science assessments that are currently used in some states, as well as time for state science instruction and assessment specialists to discuss the assessment challenges associated with the NGSS. To organize this part of the workshop, we coordinated our plans with David Heil and Sasha Burchuk, the State Collaborative on Assessment and Student Standards (SCASS) in science. The SCASS is supported by the Council of Chief State School Officers (CCSSO) and includes 29 science instruction and assessment experts in 16 states. David and Sasha were key in arranging for the wide participation of those experts in the workshop and helped us select SCASS members to serve on workshop panels. We are very grateful for the time, effort, and insights David and Sasha contributed toward making the workshop a success. We also thank the CCSSO for covering their financial contribution for the workshop.

We offer appreciation to all the state science instruction and assessment specialists who made presentations at the workshop, including Robin Anglin, West Virginia Department of Education; Anita Bernhardt, Maine Department of Education; Melinda Curless, Kentucky Department of Education; Jeff Greig, Connecticut State Department of Education; Susan Codere Kelly, Michigan Department of Education; Matt Krehbiel, Kansas State Department of Education; Shelley Lee, Wisconsin Department of Public Instruction; Yvette McCulley, Iowa Department of Education; Beverly Vance, North Carolina Department of Public Instruction; and James Woodland, Nebraska Department of Education. Their information and insights were very helpful to the committee.

After the workshop, we followed up with state instruction and assessment specialists to learn more about their science assessments. These conversations provided a great deal of background information, and we are grateful for the information and insights we received. We thank Rachel Aazzerah, Oregon Department of Education; Catherine Bowler, Massachusetts Department of Education; Liz Butner, Connecticut Department of Education; Dawn Cameron, Minnesota Department of Education; Gail Hall, Vermont Department of Education; Saundra Hamon, Kentucky Department of Education; Lauren Monowar-Jones, Ohio

Department of Education; Judy Pinnsonault, New York State Department of Education; and Brad Talbert, Utah Department of Education.

The report includes numerous examples of assessment tasks that measure science learning as envisioned in the framework and the NGSS, most of which were originally developed by committee members. Three of these examples were developed by scholars outside of the committee: Geneva Haertel and Daisy Rutstein with SRI International; Thomas Matts and Trevor Packer with the Advanced Placement Program at the College Board; and Edys Quellmalz with WestEd. We thank them for their generosity in allowing us to use their examples.

We are especially indebted to Stephen Pruitt with Achieve, Inc., who coordinated the efforts to develop the NGSS. Stephen provided us with ongoing information about the development of the standards and answered all of our questions. We sincerely appreciate his responsiveness.

The committee gratefully acknowledges the dedicated effort provided by the staff of the Board on Testing and Assessment (BOTA) and the Board on Science Education (BOSE), who worked directly on this project. Stuart Elliott, director of BOTA, and Martin Storksdieck, director of BOSE, provided leadership in moving this project forward, and their insights and guidance throughout the course of the study were invaluable. We thank Heidi Schweingruber of BOSE for her insights about the NGSS and the implications for instruction and assessment. The committee also thanks Kelly Arrington, senior project assistant, for her exceptional organizational skills and her close attention to detail. Kelly handled all of the administrative details associated with four committee meetings, held in a variety of locations, and a workshop attended by more than 100 people, and she provided critical support in preparing the manuscript.

Most especially, we express our appreciation for the extraordinary work done by Judy Koenig and Alix Beatty of BOTA in assembling critical information and in the drafting and editing of this report. Their efforts enabled the committee to push forward and meet multiple challenges related to project timelines, as well as the challenges of substantive issues regarding the design and use of educational assessments in general and for science in particular.

We also thank members of the Office of Reports and Communication of the Division of Behavioral and Social Sciences for their dedicated work on this report. We are indebted to Eugenia Grohman for her sage advice in editing numerous versions of this manuscript. We thank Kirsten Sampson Snyder for her work in coordinating a very intense review process and Yvonne Wise for shepherding the manuscript through myriad stages of production.

This report has been reviewed in draft form by individuals chosen for their diverse perspectives and technical expertise, in accordance with procedures approved by the NRC's Report Review Committee. The purpose of this independent review is to provide candid and critical comments that will assist the institution in making its published report as sound as possible and to ensure that the report meets institutional standards for objectivity, evidence, and responsiveness to the charge. The review comments and draft manuscript remain confidential to protect the integrity of the process.

We thank the following individuals for their review of this report: Charles W. (Andy) Anderson, Department of Teacher Education, Michigan State University; William B. Bridges, Department of Engineering (emeritus), California Institute of Technology; Derek Briggs, Research and Evaluation Methodology, School of Education, University of Colorado at Boulder; Angela DeBarger, Center for Technology in Learning, SRI International; George DeBoer, Project 2061, American Association for the Advancement of Science; Richard Duran, School of Education, University of California, Santa Barbara; Sean Elkins, Science Academic Program Consultant, Kentucky Department of Education; Brian Gong, Center for Assessment, Dover, New Hampshire; David Klahr, Department of Psychology, Carnegie Mellon University; Matt Krehbiel, Science Education Program, Kansas State Department of Education; Peter Labudde, Centre of Science and Technology Education, University of Applied Sciences and Arts Northwestern Switzerland; Richard C. Larson, Engineering Systems Division, Massachusetts Institute of Technology; Steven Long, Science Department, Rogers High School, Rogers, Arizona; Karen Mitchell, Association of American Medical Colleges; Mark D. Reckase, Department of Counseling, Educational Psychology and Special Education, Michigan State University; Eugenie C. Scott, National Center on Science Education, Oakland, California; Lorrie A. Shepard, School of Education, University of Colorado at Boulder; and Rebecca Zwick, Gevirtz Graduate School of Education, University of California, Santa Barbara, and Statistical Analysis, Data Analysis, and Psychometric Research, Educational Testing Service.

Although the reviewers listed above provided many constructive comments and suggestions, they were not asked to endorse the content of the report nor did they see the final draft of the report before its release. The review of this report was overseen by Lauress Wise, with HumRRO, and May Berenbaum, with the University of Illinois at Urbana-Champaign. Appointed by the NRC, they were responsible for making certain that an independent examination of this report was carried out in accordance with institutional procedures and that all review comments were carefully considered. Responsibility for the final content of this report rests entirely with the committee and the institution.

Finally, as cochairs of the committee, we thank all our fellow committee members for their dedication and outstanding contributions to this project. They actively assisted in all stages of this project, including planning the public workshop and making presentations, selecting and developing examples of assessment tasks, and writing and rewriting multiple drafts of this report. Their contributions during the late stages of the report's development, when sections of the report had to be revised on very tight schedules, are especially appreciated. They gave generously of their time and intellects throughout the project. We believe their contributions ensure that the final product is understandable to a variety of audiences and fully portrays the complex issues associated with developing the new science assessments that will be needed.

James W. Pellegrino and Mark R. Wilson, *Cochairs*
Committee on Developing Assessments of Science Proficiency in K-12

CONTENTS

SUMMARY

Science education is facing dramatic change. The new *A Framework for K-12 Science Education: Practices, Crosscutting Concepts, and Core Ideas* (hereafter, referred to as "the framework") and the *Next Generation Science Standards: For States, By States* are designed to guide educators in significantly altering the way science is taught—from kindergarten through high school (K-12). The framework is aimed at making science education more closely resemble the way scientists actually work and think. It is also aimed at making instruction reflect research on learning that demonstrates the importance of building coherent understandings over time.

The framework structures science learning around three dimensions: the *practices* through which scientists and engineers do their work; the key *crosscutting concepts* that link the science disciplines; and the *core ideas* of the disciplines of life sciences, physical sciences, earth and space sciences, and engineering and technology. It argues that they should be interwoven in every aspect of science education, most critically, curriculum, instruction, and assessment. The framework emphasizes the importance of the connections among the disciplinary core ideas, such as using understandings about chemical interactions from physical science to explain biological phenomena.

We use the term "three-dimensional science learning" to refer to the integration of these dimensions. It describes not the process of learning, but the kind of thinking and understanding that science education should foster. The framework and the Next Generation Science Standards (NGSS) are also grounded in the ideas

1

that science learning develops over time and assessments will need to mark students' progress toward specific learning objectives.

This new vision of science learning presents considerable challenges—but also a unique and valuable opportunity for assessment. Existing science assessments have not been designed to capture three-dimensional science learning, and developing assessments that can do so requires new approaches. Rethinking science assessment in this way also offers an opportunity to address long-standing problems with current approaches. In this context, the following charge was given to the Committee on Developing Assessments of Science Proficiency in K-12:

The committee will make recommendations for strategies for developing assessments that validly measure student proficiency in science as laid out in the new K-12 science education framework. The committee will review recent and current, ongoing work in science assessment to determine which aspects of the necessary assessment system for the framework's vision can be assessed with available techniques and what additional research and development is required to create an overall assessment system for science education in K-12. The committee will prepare a report that includes a conceptual framework for science assessment in K-12 and will make recommendations to state and national policy makers, research organizations, assessment developers, and study sponsors about the steps needed to develop valid, reliable, and fair assessments for the framework's vision of science education. The committee's report will discuss the feasibility and cost of its recommendations.

AN ASSESSMENT SYSTEM

The NGSS describe specific goals for science learning in the form of *performance expectations*—statements about what students should know and be able to do at each grade level—and thus what should be tested at each grade level. Each performance expectation incorporates all three dimensions, and the NGSS emphasize the importance of the connections among scientific concepts. The NGSS's performance expectations place significant demands on science learning at every grade level. It will not be feasible to assess all of the performance expectations for a given grade level during a single assessment occasion. Students will need multiple—and varied—assessment opportunities to demonstrate their competence on the performance expectations for a given grade level (Conclusion 2-3).[1]

---

[1]The conclusion and recommendation numbers refer to the report's chapters and the order in which they appear.

In addition, the effective evaluation of three-dimensional science learning will require more than a one-to-one mapping between the performance expectations and assessment tasks. More than one assessment task may be needed to adequately assess students' mastery of some performance expectations, and any given assessment task may assess aspects of more than one performance expectation. Moreover, to assess both understanding of core knowledge and facility with a practice, assessments may need to probe students' use of a given practice in more than one disciplinary context. Assessment tasks that attempt to test practices in isolation from one another may not be meaningful as assessments of the three-dimensional science learning called for by the NGSS (Conclusion 2-4).

To adequately cover the three dimensions, assessment tasks will need to contain multiple components, such as a set of interrelated questions. It may be useful to focus on individual practices, core ideas, or crosscutting concepts in a specific component of an assessment task, but, together, the components need to support inferences about students' three-dimensional science learning as described in a given performance expectation (Conclusion 2-1).

Measuring the learning described in the NGSS will require assessments that are significantly different from those in current use. Specifically, the tasks designed to assess the performance expectations in the NGSS will need to have the following characteristics (Conclusion 4-1):

- include multiple components that reflect the connected use of different scientific practices in the context of interconnected disciplinary ideas and crosscutting concepts;
- address the progressive nature of learning by providing information about where students fall on a continuum between expected beginning and ending points in a given unit or grade; and
- include an interpretive system for evaluating a range of student products that are specific enough to be useful for helping teachers understand the range of student responses and provide tools for helping teachers decide on next steps in instruction.

Designing specific assessment tasks and assembling them into tests will require a careful approach to assessment design. Some currently used approaches, such as evidence-centered design and construct modeling, do reflect such design through the use of the fundamentals of cognitive research and theory. With these approaches, the selection and development of assessment tasks, as well as the scoring rubrics and criteria for scoring, are guided by the construct to be assessed and the best ways of eliciting evidence about students' proficiency with that construct. In designing assessments to measure proficiency on the NGSS performance expectations, the committee recommends the use of one of these approaches (Recommendation 3-1).

More broadly, a system of assessments will be needed to measure the NGSS performance expectations and provide students, teachers, administrators, policy makers, and the public with the information each needs about student learning (Conclusion 6-1). This conclusion builds on the advice in prior reports of the National Research Council. We envision a range of assessment strategies that are designed to answer different kinds of questions with appropriate degrees of specificity and provide results that complement one another. Such a system needs to include three components:

1. assessments designed to support classroom instruction,
2. assessments designed to monitor science learning on a broader scale, and
3. a series of indicators to monitor that the students are provided with adequate opportunity to learn science in the ways laid out in the framework and the NGSS.

CLASSROOM ASSESSMENTS

Classroom assessments are an integral part of instruction and learning and should include both formative and summative tasks: formative tasks are those that are specifically designed to be used to guide instructional decision making and lesson planning; summative tasks are those that are specifically designed to assign student grades.

The kind of instruction that will be effective in teaching science in the way the framework and the NGSS envision will require students to engage in scientific and engineering practices in the context of disciplinary core ideas—and to make connections across topics through the crosscutting ideas. To develop the skills and dispositions to use scientific and engineering practices needed to further their learning and to solve problems, students need to experience instruction in which they (1) use multiple practices in developing a particular core idea and (2) apply each practice in the context of multiple core ideas. Effective use of the practices often requires that they be used in concert with one another, such as in supporting explanation with an argument or using mathematics to analyze data (Conclusion 4-2).

Assessment activities will be critical supports for this instruction. Students will need guidance about what is expected of them and opportunities to reflect on their performance as they develop proficiencies. Teachers will need information about what students understand and can do so they can adapt their instruction. Instruction that is aligned with the framework and the NGSS will naturally provide many opportunities for teachers to observe and record evidence of students learning. The student activities that reflect such learning include developing and refining models; generating, discussing, and analyzing data; engaging in both spoken and written explanations and argumentation; and reflecting on their own understanding. Such opportunities are the basis for the development of assessments of three-dimensional science learning.

Assessment tasks that have been designed to be integral to classroom instruction—in which the kinds of activities that are part of high-quality instruction are deployed in particular ways to yield assessment information—are beginning to be developed. They demonstrate that it is possible to design tasks that elicit students' thinking about disciplinary core ideas and crosscutting concepts by engaging them in scientific practices and that students can respond to them successfully (Conclusion 4-3). However, these types of assessments of three-dimensional science learning are challenging to design, implement, and properly interpret. Teachers will need extensive professional development to successfully incorporate this type of assessment into their practice (Conclusion 4-4).

State and district leaders who design professional development for teachers should ensure that it addresses the changes posed by the framework and the NGSS in both the design and use of assessment tasks as well as instructional strategies. Professional development has to support teachers in integrating practices, crosscutting concepts, and disciplinary core ideas in inclusive and engaging instruction and in using new modes of assessment that support such instructional activities (Recommendation 4-1).

Curriculum developers, assessment developers, and others who create instructional units and resource materials aligned to the new science framework and the NGSS will need to ensure that assessment activities included in such materials (such as mid- and end-of-chapter activities, suggested tasks for unit assessment, and online activities) require students to engage in practices that demonstrate their understanding of core ideas and crosscutting concepts. These materials should also attend to multiple dimensions of diversity (e.g., by connecting with students' cultural and linguistic resources). In designing these materials, development teams need to include experts in science, science learning, assessment design, equity and diversity, and science teaching (Recommendation 4-2).

MONITORING ASSESSMENTS

Assessments designed for monitoring purposes, also referred to as external assessments, are used to audit student learning over time. They are used to answer important questions about student learning such as: How much have the students in a certain school system learned over the course of a year? How does achievement in one school system compare with achievement in another? Is one instructional technique or curricular program more effective than another? What are the effects of a particular policy measure such as reduction in class size?

To measure the NGSS performance expectations, the tasks used in assessments designed for monitoring purposes need to have the same characteristics as those used for classroom assessments. But assessments used for monitoring pose additional challenges: they need to be designed so that they can be given to large numbers of students, to be sufficiently standardized to support the intended monitoring purpose, to cover an appropriate breadth of the NGSS, and to be feasible and cost-effective for states.

The multicomponent tasks needed to effectively evaluate the NGSS performance expectations will include a variety of response formats, including performance-based questions, those that require students to construct or supply an answer, produce a product, or perform an activity. Although performance-based questions are especially suitable for assessing some aspects of student proficiency on the NGSS performance expectations, it will not be feasible to cover the full breadth and depth of the NGSS performance expectations for a given grade level with a single external assessment comprised solely or mostly of performance-based questions: performance-based questions take too much time to complete, and many of them would be needed in order to fully cover the set of performance expectations for a grade level. Consequently, the information from external on-demand assessments (i.e., assessments that are administered at a time mandated by the state) will need to be supplemented with information gathered from classroom-embedded assessments (i.e., assessments that are administered at a time determined by the district or school that fits the instructional sequence in the classroom) to fully cover the breadth and depth of the performance expectations. Both kinds of assessments will need to be designed so that they produce information that is appropriate and valid to support a specific monitoring purpose (Recommendation 6-1).

Classroom-embedded assessments may take various forms. They could be self-contained curricular units, which include instructional materials and assessments provided by the state or district to be administered in classrooms. Alternatively, a state or district might develop item banks of tasks that could be used at the appropriate time in classrooms. States or districts might require that students in certain grade levels assemble portfolios of work products that demonstrate their levels of proficiency. Using classroom-embedded assessments for monitoring purposes leaves a number of important decisions to the district or school; quality control procedures would need to be implemented so that these assessments meet appropriate technical standards (Conclusion 5-2).

[handwritten margin note: Can we develop along w/ science Magnets for the district?]

External assessments would consist of sets of multicomponent tasks. To the extent possible, these tasks should include—as a significant and visible aspect of the assessment—multiple, performance-based questions. When appropriate, computer-based technology should be used to broaden and deepen the range of performances used on these assessments (Recommendation 6-2).

Assessments that include performance-based questions can pose technical and practical challenges for some monitoring purposes. For instance, it can be difficult both to attain appropriate levels of score reliability and to produce results that can be compared across groups or across time, comparisons that are important for monitoring. Developing, administering, and scoring the tasks can be time consuming and resource intensive. To help address these challenges, assessment developers should take advantage of emerging and validated innovations in assessment design, scoring, and reporting to create and implement assessments of three-dimensional science learning (Recommendation 5-2). In particular, state and local policy makers should design the external assessment component of their systems so that they incorporate the use of matrix-sampling designs whenever appropriate (rather than requiring that every student take every item).

INDICATORS OF OPPORTUNITY TO LEARN

Indicators of the opportunity to learn make it possible to evaluate the effectiveness of science instructional programs and the equity of students' opportunity to learn science in the ways envisioned by the new framework. States should routinely collect information to monitor the quality of the classroom instruction in science, the extent to which students have the opportunity to learn science in the way called for in the framework, and the extent to which schools have the resources needed to support learning (such as teacher qualification and subject area pedagogical knowledge, and time, space, and materials devoted to science instruction) (Recommendation 6-6).

Measures of the quality and content of instruction should also cover inclusive instructional approaches that reach students of varying cultural and linguistic backgrounds. Because assessment results cannot be fully understood in the absence

Our school portfolio →

of information about the opportunities to learn what is tested, states should collect relevant indicators—including the material, human, and social resources available—to support student learning so they can contextualize and validate the inferences drawn from assessment results (Recommendation 7-6). This information should be collected through inspections of school science programs, surveys of students and teachers, monitoring of teacher professional development programs, and documentation of curriculum assignments and student work.

IMPLEMENTATION

The assessment system that the committee recommends differs markedly from current practice and will thus take time to implement, just as it will take time to adopt the instructional programs needed for students to learn science in the way envisioned in the framework and the NGSS. States should develop and implement new assessment systems gradually and establish carefully considered priorities. Those priorities should begin with what is both necessary and possible in the short term while also establishing long-term goals leading to implementation of a fully integrated and coherent system of curriculum, instruction, and assessment (Recommendation 7-1).

The committee encourages a developmental path for assessment that is "bottom up" rather than "top down": one that begins with the process of designing assessments for the classroom, perhaps integrated into instructional units, and moves toward assessments for monitoring. In designing and implementing their assessment systems, states will need to focus on professional development. States will need to include adequate time and resources for professional development so that teachers can be properly prepared and guided and so that curriculum and assessment developers can adapt their work to the vision of the framework and the NGSS (Recommendation 7-2).

State and district leaders who commission assessment development should ensure that the plans address the changes called for by the framework and the NGSS. They should build into their commissions adequate provision for the substantial amounts of time, effort, and refinement that are needed to develop and implement the use of such assessments: multiple cycles of design-based research will be necessary (Recommendation 7-3).

Existing and emerging technologies will be critical tools for creating a science assessment system that meets the goals of the framework and the NGSS, particularly those that permit the assessment of three-dimensional knowledge, as well as the streamlining of assessment administration and scoring

Measurement for assessment development?

(Recommendation 7-7). States will be able to capitalize on efforts already under way to implement the new Common Core State Standards in English language arts and mathematics, which have required educators to integrate learning expectations and instruction. Nevertheless, the approach to science assessment that the committee recommends will still require modifications to current systems. States will need to carefully lay out their priorities and adopt a thoughtful, reflective, and gradual process for making the transition to an assessment system that supports the vision of the framework and the NGSS.

A fundamental component of the framework's vision for science education is that all students can attain its learning goals. The framework and the NGSS both stress that this can only happen if all students have the opportunity to learn in the new ways called for and if science educators are trained to work with multiple dimensions of diversity. A good assessment system can play a critical role in providing fair and accurate measures of the learning of all students and providing students with multiple ways of demonstrating their competency. Such an assessment system will include formats and presentation of tasks and scoring procedures that reflect multiple dimensions of diversity, including culture, language, ethnicity, gender, and disability. Individuals with expertise in diversity should be integral participants in developing state assessment systems (Recommendation 7-5).

1

INTRODUCTION

A Framework for K-12 Science Education: Practices, Crosscutting Concepts, and Core Ideas (National Research Council, 2012a, hereafter referred to as "the framework") provided the foundation for new science education standards, which were published the following year (NGSS Lead States, 2013). The framework is grounded in a new vision for science education from kindergarten through high school (K-12): that all students—not just those who intend to pursue science beyond high school—will learn core scientific ideas in increasing depth over multiple years of schooling. It calls for an approach to education that closely mirrors the way that science is practiced and applied, and it focuses on the cumulative learning opportunities needed to ensure that (National Research Council, 2012a, p. 1):

[By] the end of 12th grade, all students have some appreciation of the beauty and wonder of science; possess sufficient knowledge of science and engineering to engage in public discussions on related issues; are careful consumers of scientific and technological information related to their everyday lives; are able to continue to learn about science outside school; and have the skills to enter careers of their choice, including (but not limited to) careers in science, engineering, and technology.

The framework cites well-known limitations in K-12 science education in the United States—that it "is not organized systematically across multiple years of school, emphasizes discrete facts with a focus on breadth over depth, and does not provide students with engaging opportunities to experience how science is actually done" (p. 1). To address these limitations, the framework details three dimensions for science education—the *practices* through which scientists and engineers do their work, the key *crosscutting concepts* for all disciplines, and the *core ideas* of the disciplines—and it argues that the dimensions need to be interwoven in every aspect of science education, including assessment.

Developing new assessments to measure the kinds of learning the framework describes presents a significant challenge and will require a major change to the status quo. The framework calls for assessments that capture students' competencies in performing the practices of science and engineering by applying the knowledge and skills they have learned. The assessments that are now in wide use were not designed to meet this vision of science proficiency and cannot readily be retrofitted to do so. To address this disjuncture, the Committee on Developing Assessments of Science Proficiency in K-12 was asked to help guide the development of new science assessments.

The committee was charged to make recommendations to state and national policy makers, research organizations, assessment developers, and funders about ways to use best practices to develop effective, fair, reliable, and high-quality assessment systems that support valid conclusions about student learning. The committee was asked to review current assessment approaches and promising research and to develop both a conceptual framework for K-12 science assessment and an analysis of feasibility issues. The committee's full charge is shown in Box 1-1.

CONTEXT

Science education has been under a great deal of scrutiny for several decades. Policy makers have lamented that the United States is falling behind in science, technology, engineering, and mathematics (STEM) education, based on international comparisons and on complaints that U.S. students are not well prepared for the workforce of the 21st century (see, e.g., National Research Council, 2007). The fact that women and some demographic groups are significantly underrepresented in postsecondary STEM education and in STEM careers is another fact that has captured attention (Bystydzienski and Bird, 2006; National Research Council, 2011a; Burke and Mattis, 2007). The framework discusses ways in which some student groups have been excluded from science and the need to better link science instruction to diverse students' interests and experiences.[1]

---

[1]See Chapter 11 of the framework (National Research Council, 2012a) for discussion of these issues.

Researchers, educators, and others have argued that a primary reason for the problems is the way science is taught in U.S. schools (see, e.g., National Research Council, 2006; National Task Force on Teacher Education in Physics, 2010; Davis et al., 2006; Association of Public and Land-grant Universities, 2011). They have pointed out specific challenges—for example, that many teachers who are responsible for science have not been provided with the knowledge and skills required to teach in the discipline they are teaching or in science education[2]— and the lack of adequate instructional time and adequate space and equipment for investigation and experimentation in many schools (OECD, 2011; National Research Council, 2005). Another key issue has been the inequity in access to instructional time on science and associated resources and its influence on the performance of different demographic groups of students. Others have focused on a broader failing, arguing that K-12 science education is generally too disconnected from the way science and engineering are practiced and should be reformed. The framework reflects and incorporates these perspectives.

The framework's approach is also grounded in a growing body of research on how young people learn science, which is relevant to both instruction and

---

[2]This critique is generally targeted to both middle and secondary school teachers, who are usually science specialists, and elementary teachers who are responsible for teaching several subjects.

assessment. Researchers and practitioners have built an increasingly compelling picture of the cumulative development of conceptual understanding and the importance of instruction that guides students in a coherent way across the grades (National Research Council, 2001, 2006). A related line of research has focused on the importance of instruction that is accessible to students of different backgrounds and uses their varied experiences as a base on which to build. These newer models of how students learn science are increasingly dominant in the science education community, but feasible means of widely implementing changes in teacher practice that capitalize on these ideas have been emerging only gradually.

The new framework builds on influential documents about science education for K-12 students, including the *National Science Education Standards* (National Research Council, 1996) the *Benchmarks for Science Literacy: A Tool for Curriculum Reform* (American Association for the Advancement of Science, 1993, 2009), and the *Science Framework for the 2009 National Assessment of Educational Progress* (National Assessment Governing Board, 2009). At the same time, the landscape of academic standards has changed significantly in the last few years, as the majority of states have agreed to adopt common standards in language arts and mathematics.[3]

National and state assessment programs, as well as international ones, have been exploring new directions in assessment and will be useful examples for the developers of new science assessments. Two multistate consortia received grants under the federal Race to the Top Assessment Program to develop innovative assessment in language arts and mathematics that will align with the new Common Core State Standards. The Partnership for Assessment of Readiness for College and Careers (PARCC) and the Smarter Balanced Assessment Consortium (SBAC) are working to develop assessments that can be implemented during the 2014-2015 school year.[4] We have followed their progress closely, but our recommendations for science assessment are completely separate from their work. Examples from international science assessments and the approach to developing assessments for the revised Advanced Placement Program in high schools in biology are other valuable models.

---

[3]The Common Core State Standards have been adopted by 45 states, the District of Columbia, four territories, and the U.S. Department of Defense Education Activity. For more information, see http://www.corestandards.org/ [August 2013].

[4]For details, see http://www2.ed.gov/programs/racetothetop/index.html [June 2013]. Information about PARCC, SBAC, and the Common Core State Standards can be found, respectively, at http://www.parcconline.org/about-parcc, http://www.smarterbalanced.org/, and http://www.corestandards.org/ [June 2013].

New standards, called the Next Generation Science Standards (NGSS), have been developed specifically in response to the approach laid out in the framework by a team of 26 states that are working with Achieve, Inc. The developers included representatives from science, engineering, science education, higher education, and business and industry (NGSS Lead States, 2013). Draft versions of the document were subjected to revisions based on extensive feedback from stakeholders and two rounds of public comment. The NGSS team also worked to coordinate the new science standards with new Common Core State Standards in English language arts and mathematics so that intellectual links among the disciplines can be emphasized in instruction. Preliminary drafts were available in May 2012, January 2013, and the final version of the NGSS was released in April 2013.

NEED FOR FUNDAMENTAL CHANGE

The new K-12 science framework provides an opportunity to rethink the role that assessment plays in science education. The most fundamental change the framework advocates—that understanding of core ideas and crosscutting concepts be completely integrated with the practices of science—requires changes in the expectations for science assessment and in the nature of the assessments used.

At present, the primary purpose of state-level assessment in the United States is to provide information that can be used for accountability purposes. Most states have responded to the requirements of the No Child Left Behind Act of 2001 (NCLB) by focusing their assessment resources on a narrow range of assessment goals. In science, NCLB requires formal, statewide assessment once in three clusters of grades (3-5, 6-9, and high school).[5] That is, unless a state does more than NCLB requires, students' understanding of science is formally evaluated only three times from kindergarten through grade 12, usually with state assessments that are centrally designed and administered. This approach to assessment does not align with the goals of the new framework: it does not reflect the importance of students' gradual progress toward learning goals. Monitoring of student learning is important, but most current tests do not require students to demonstrate knowledge of the integration between scientific practices and conceptual understanding. The NGSS, for example, include an expectation that students understand how the way in which scientific phenomena are modeled may influence conceptual under-

---

[5]NCLB requires testing in mathematics and language arts every year.

standing, but few current science assessments evaluate this aspect of science. Thus, aligning new tests with the framework's structure and goals will require the use of a range of assessment tools designed to meet a variety of needs for information about how well students are learning complex concepts and practices.

Among the states, the time, resources, and requirements for testing students in science vary widely: states each have devised their own combination of grades tested, subject areas covered, testing formats, and reporting strategies. Most states rely heavily on assessments that are affordable, efficient, and easily standardized: these are generally easy-to-score multiple-choice and short open-ended questions that assess recall of facts. Assessments used as benchmarks of progress, and even those embedded in curriculum, often use basic and efficient paper-and-pencil formats.

Although the various state science assessments often provide technically valid and reliable information for specific purposes, they cannot systematically assess the learning described in the framework and the three-dimensional performance standards described in the NGSS. New kinds of science assessments are needed to support the new vision and understanding of students' science learning. Developing an assessment program that meets these new goals presents complex conceptual, technical, and practical challenges, including cost and efficiency, obtaining reliable results from new assessment types, and developing complex tasks that are equitable for students across a wide range of demographic characteristics.

COMMITTEE'S APPROACH

The committee's charge led us first to a detailed review of what is called for by the framework and the NGSS. We were not asked to take a position on these documents. The framework sets forth goals for science learning for all students that will require significant shifts in curriculum, instruction, and assessment. The NGSS represent a substantial and credible effort to map the complex, three-dimensional structure of the framework into a coherent set of performance expectations to guide the development of assessments (as well as curriculum and instruction). The committee recognizes that some mapping of this kind is an essential step in the alignment of assessments to the framework, and the NGSS are an excellent beginning. We frequently consulted both documents: the framework for the vision of student learning and the NGSS for specific characterization of the types of outcomes that will be expected of students.

We also examined prior National Research Council reports, such as *Knowing What Students Know* (National Research Council, 2001) and *Systems for State Science Assessment* (National Research Council, 2006), and other materials that are relevant to the systems approach to assessment called for in the new framework. And we explored research and practice in educational measurement that are relevant to our charge: the kinds of information that can be obtained using large-scale assessments; the potential benefits made possible by technological and other innovations; what can be learned from recent examples of new approaches, including those used outside the United States; and the results of attempts to implement performance assessments as part of education reform measures in the 1980s and 1990s. Last, we examined research and practice related to classroom-based assessments in science and the role of learning progressions in guiding approaches to science curricula, instruction, and assessment.

As noted above, this project was carried out in the context of developments that in many cases are rapidly altering the education landscape. The committee devoted attention to tracking the development of the NGSS and the implementation of the new Common Core State Standards.[6] As this report went to press, 11 states and the District of Columbia had adopted the NGSS.[7] The work of PARCC and SBAC, which are developing assessments to align with the Common Core State Standards and have explored some current technological possibilities, has also been important for the committee to track. However, we note that both consortia were constrained in their decisions about technology and task design, both by the challenge of assessing every student every year, as mandated by NCLB for mathematics and language arts, and by a timeline for full implementation that left little space for exploration of some of the more innovative options that we explored for science.

This committee's charge required a somewhat unusual approach. Most National Research Council committees rely primarily on syntheses of the research literature in areas related to their charge as the basis for their conclusions and recommendations. However, the approach to instruction and assessment envisioned in the framework and the NGSS is new: thus, there is little research on which to base our recommendations for best strategies for assessment. Furthermore, the

---

[6]For details, see http://www.corestandards.org/in-the-states [June 2013].

[7]As of April 2014, the states were California, Delaware, Illinois, Kansas, Kentucky, Maryland, Nevada, Oregon, Rhode Island, Vermont, Washington, and the District of Columbia.

development of the NGSS occurred while our work was underway, and so we did not have the benefit of the final version until our work was nearly finished.

In carrying out our charge, we did review the available research in relevant fields, including educational measurement, cognitive science, learning sciences, and science education, and our recommendations are grounded in that research. They are also the product of our collective judgment about the most promising ways to make use of tools and ideas that are already familiar, as well as our collective judgment about some tools and ideas that are new, at least for large-scale applications in the United States. Our charge required that we consider very recent research and practice, alongside more established bodies of work, and to develop actionable recommendations on the basis of our findings and judgments. We believe our recommendations for science assessment can be implemented to support the changes called for in the framework.

Much of our research focused on gathering information on the types of science assessments that states currently use and the types of innovations that might be feasible in the near future. We considered this information in light of new assessment strategies that states will be using as part of their efforts to develop language arts and mathematics assessments for the Common Core through the Race to the Top consortia, particularly assessments that make use of constructed-response and performance-based tasks and technology-enhanced questions. To help us learn more about these efforts, representatives from the two consortia (PARCC and SBAC) made presentations at our first meeting, and several committee members participated in the June 2012 Invitational Research Symposium on Technology Enhanced Assessments, sponsored by the K-12 Center at the Educational Testing Service. That symposium focused on the types of innovations under consideration for use with the consortia-developed assessments, including the use of technology to assess hard-to-measure constructs and expand accessibility, the use of such innovative formats as simulations and games, and the development of embedded assessments.

We took a number of other steps to learn more about states' science assessments. We reviewed data from a survey of states conducted by the Council of State Science Supervisors on the science assessments they used in 2012, the grades they tested, and the types of questions they used. Based on these survey data, we identified states that made use of any types of open-ended questions, performance tasks, or technology enhancements and followed up with the science specialists in those states: Massachusetts, Minnesota, New Hampshire, New York, Ohio, Oregon, Rhode Island, Vermont, and Utah.

During the course of our data gathering, a science assessment specialist in Massachusetts organized a webinar on states' efforts to develop performance-based tasks. Through this webinar we learned of work under way in Connecticut, Ohio, and Vermont. Members of the committee also attended meetings of the State Collaborative on Assessment and Student Standards (SCASS) in science of the Council of Chief State School Officers (CCSSO) and a conference on building capacity for state science education sponsored by the CCSSO and the Council of State Science Supervisors.

We also held a public workshop, which we organized in conjunction with the SCASS. The workshop included presentations on a range of innovative assessments, including the College Board's redesigned Advanced Placement Biology Program, the 2009 science assessment by the National Assessment of Educational Progress that made use of computer-interactive and hands-on tasks, WestEd's SimScientist Program, and curriculum-embedded assessments from the middle school curriculum materials of IQWST (Investigating and Questioning our World through Science and Technology, Krajcik et al., 2013). SCASS members served as discussants at the workshop. The workshop was an opportunity to hear from researchers and practitioners about their perspectives on the challenges and possibilities for assessing science learning, as well as to hear about various state assessment programs. The workshop agenda appears in Appendix A.

GUIDE TO THE REPORT

Throughout the report the committee offers examples of assessment tasks that embody our approach and demonstrate what we think will be needed to measure science learning as described in the framework and the NGSS. Because the final version of the NGSS was not available until we had nearly completed work on this report, none of the examples was specifically aligned with the NGSS performance expectations. However, the examples reflect the ideas about teaching, learning, and assessment that influenced the framework and the NGSS, and they can serve as models of assessment tasks that measure both science content and practice.[8] The examples have all been used in practice and appear in Chapters 2, 3, 4, and 5: see Table 1-1 for a summary of the example tasks included in the

---

[8]These examples were developed by committee members and other researchers prior to this study.

TABLE 1-1 Guide to Examples of Assessment Tasks in the Report

Chapter and Example	Disciplinary Core Idea[a]	Practices	Crosscutting Concepts	Grade Level
1 What Is Going on Inside Me? (Chapter 2)	PS1: Matter and its interactions LS1: From molecules to organisms: Structures and processes	Constructing explanations Engaging in argument from evidence	Energy and matter: flows, cycles, and conservation	Middle school
2 Pinball Car (Chapter 3)	PS3: Energy	Planning and carrying out investigations	Energy and matter: flows, cycles, and conservation	Middle school
✓ 3 Measuring Silkworms (Chapters 3 and 4)	LS1.A: Structure and function: Organisms have macroscopic structures that allow for growth LS1.B Growth and development of organisms: Organisms have unique and diverse life cycles	Asking questions Planning and carrying out investigations Analyzing and interpreting data Using mathematics Constructing explanations Engaging in argument from evidence Communicating information	Patterns	✓ Grade 3
4 Behavior of Air (Chapter 4)	PS1: Matter and its interactions	Developing and using models Engaging in argument from evidence	Energy and matter: flows, cycles, and conservation. Systems and system models	Middle school
✓ 6 Biodiversity in the Schoolyard (Chapter 4)	LS4: Biological evolution: Unity and diversity	Planning and carrying out investigations[b] Analyzing and interpreting data Constructing explanations	Patterns	✓ Grade 5
7 Climate Change (Chapter 4)	LS2: Ecosystems: Interactions, energy, and dynamics ESS3-5: Earth and human activity	Analyzing and interpreting data Using a model to predict phenomena	System and system models	High school

TABLE 1-1 Continued

Chapter and Example	Disciplinary Core Idea[a]	Practices	Crosscutting Concepts	Grade Level
8 Ecosystems (Chapter 4)	LS2: Ecosystems: Interactions, energy, and dynamics	Planning and carrying out investigations and interpreting patterns	Systems and system models Patterns	
9 Photosyntheses and Plant Evolution (Chapter 5)	LS4: Biological evolution: Unity and diversity	Developing and using models Analyzing and interpreting data Using mathematics and computational thinking Constructing explanations	Systems and system models Patterns	High school
10 Sinking and Floating (Chapter 5)	PS2: Motion and stability	Obtaining, evaluating, and communicating information Asking questions Planning and carrying out investigations Analyzing and interpreting data Engaging in argument from evidence	Cause and effect Stability and change	Grade 2
11 Plate Tectonics (Chapter 5)	ESS2: Earth's systems	Developing and using models Constructing explanations	Patterns Scale, proportion, and quantity	Middle school

[a] ESS = earth and space sciences; LS = life sciences; PS = physical sciences. The disciplinary codes are taken from the new science framework: see Box 2-1 in Chapter 2.
[b]This example focuses on carrying out an investigation.

report and the disciplinary core ideas, practices, and crosscutting concepts that they are intended to measure.

The report is structured around the steps that will be required to develop assessments to evaluate students' proficiency with the NGSS performance expectations, and we use the examples to illustrate those steps. The report begins, in Chapter 2, with an examination of what the new science framework and the NGSS require of assessments. The NGSS and framework emphasize that science learning involves the active engagement of scientific and engineering practices in the context of disciplinary core ideas and crosscutting concepts—a type of learning that we refer to as "three-dimensional learning." The first of our example assessment tasks appears in this chapter to demonstrate what three-dimensional learning involves and how it might be assessed.

Chapter 3 provides an overview of the fundamentals of assessment design. In the chapter, we discuss "principled" approaches to assessment design: they are principled in that they provide a methodical and systematic approach to designing assessment tasks that elicit performances that accurately reflect students' proficiency. We use the example assessment task in the chapter to illustrate this type of approach to developing assessments.

Chapter 4 focuses on the design of classroom assessment tasks that can measure the performance expectations in the NGSS. The chapter addresses assessment tasks that are administered in the classroom for both formative and summative purposes. We elaborate on strategies for designing assessment tasks that can be used for either of these assessment purposes, and we include examples to illustrate the strategies.

Chapter 5 moves beyond the classroom setting and focuses on assessments designed to monitor science learning across the country, such as to document students' science achievement across time; to compare student performance across schools, districts, or states; or to evaluate the effectiveness of certain curricula or instructional practices. The chapter addresses strategies for designing assessment tasks that can be administered on a large scale, such as to all students in a school, district, or state. The chapter addresses the technical measurement issues associated with designing assessments (i.e., assembling groups of tasks into tests, administering them, and scoring the responses) so that the resulting performance data provide reliable, valid, and fair information that can be used for a specific monitoring purpose.

Chapter 6 discusses approaches to developing a coherent system of curricula, instruction, and assessments that together support and evaluate students' science learning.

Finally, in Chapter 7 we address feasibility issues and explore the challenges associated with implementing the assessment strategies that we recommend. Those challenges include the central one of accurately assessing the science learning of all students, particularly while substantial change is under way. The equity issues that are part of this challenge are addressed in Chapter 7 and elsewhere in the report.

2

ASSESSMENTS TO MEET THE GOALS OF THE FRAMEWORK

The committee's charge is to recommend best practices for developing reliable and valid assessments that measure student proficiency in science as conceptualized in *A Framework for K-12 Science Education: Practices, Crosscutting Concepts, and Core Ideas* (National Research Council, 2012a, hereafter referred to as "the framework") and the *Next Generation Science Standards: For States, By States* (NGSS Lead States, 2013). In this chapter, we review the main features of these two documents with respect to the assessment challenges they pose.[1]

THE FRAMEWORK'S VISION FOR K-12 SCIENCE EDUCATION

There are four key elements of the framework's vision for science education that will likely require significant change in most science classrooms:

1. a focus on developing students' understanding of a limited set of core ideas in the disciplines and a set of crosscutting concepts that connect them;
2. an emphasis on how these core ideas develop over time as students' progress through the K-12 system and how students make connections among ideas from different disciplines;

---

[1]We refer readers to the framework and the Next Generation Science Standards for a complete picture of what they propose for science education.

3. a definition of learning as engagement in the science and engineering practices to develop, investigate, and use scientific knowledge; and

4. an assertion that science and engineering learning for all students will entail providing the requisite resources and more inclusive and motivating approaches to instruction and assessment, with specific attention to the needs of disadvantaged students.

The framework was built on previous documents that lay out expectations for K-12 learning in science, drawing on ideas developed in *National Science Education Standards* (National Research Council, 1996), the *Benchmarks for Science Literacy* (American Association for the Advancement of Science, 1993, 2009), *Science Framework for the 2009 National Assessment of Educational Progress* (National Assessment Governing Board, 2009), and the *Science College Board Standards for College Success* (College Board, 2009).

The design of the framework was also influenced by a body of research conducted over the 15 years since the publication of *National Science Education Standards* (National Research Council, 1996). This research demonstrates that science and engineering involve both knowing and doing; that developing rich, conceptual understanding is more productive for future learning than memorizing discrete facts; and that learning experiences should be designed with coherent progressions over multiple years in mind (see research syntheses in National Research Council, 2006, 2007, 2009; National Academy of Engineering and National Research Council, 2009). Thus, the goal of science education, as articulated in the framework, is to help all students consciously and continually build on and revise their knowledge and abilities through engagement in the practices of science and engineering.

The framework also emphasizes the connections among science, engineering, and technology. Key practices and ideas from engineering are included because of the interconnections between science and engineering and because there is some evidence that engaging in engineering design can help to leverage student learning in science. The goal of including ideas related to engineering, technology, and the applications of science in the framework for science education is not to change or replace current K-12 engineering and technology courses (typically offered only at the high school level as part of career and technical education offerings). Rather, the goal is to strengthen science education by helping students understand the similarities and differences between science and engineering by making the connections between the two fields explicit and by providing all students with an introduction to engineering.

The concept of equity is integral to the framework's definition of excellence. The framework's goals are explicitly intended for all students, and it emphasizes that learners from diverse backgrounds can indeed engage in and learn complex subject matter. The Next Generation Science Standards (NGSS) also highlight issues related to equity and diversity and offer specific guidance for fostering science learning for diverse groups (see NGSS Lead States, 2013, Appendix D). It notes important challenges: students' opportunities to learn are rarely equitable, and the changes to curriculum and instruction called for may take longest to reach the students already at the greatest disadvantage in science education. Opportunity to learn is a matter not only of resources, such as instructional time, equipment, and materials, and well-prepared teachers; it is also a matter of the degree to which instruction is designed to meet the needs of diverse students and to identify, draw on, and connect with the advantages their diverse experiences give them for learning science. This conception of opportunity to learn will be key to meeting the framework's vision, as it explicitly notes (NGSS Lead States, 2013, p. 28). There is increasing recognition that the diverse customs and orientations that members of different cultural communities bring to both formal and to informal science learning are assets on which to build. Teachers can connect this rich cultural base to classroom learning by embracing diversity as a means of enhancing learning about science and the world.

Although brief, the above description makes clear the extent of the challenge posed by the framework's definition of excellence. Assessment designers are faced with the challenge of finding a balance among three competing priorities: (1) using assessment as a tool for supporting and promoting an ambitious vision for all students, (2) obtaining accurate measures of what students have actually learned, and (3) supporting equity of opportunity for disadvantaged students. If the implementation of the NGSS proceeds as intended, then new assessment designs will be developed and implemented in the context of significant changes to all aspects of science education—a circumstance that magnifies the challenge of finding the right balance among the three priorities. And all of these challenges arise in the context of serving all students. The myriad issues associated with meeting these challenges and, more broadly, the framework's goals of science education for all students, are beyond the committee's charge. We do, however, highlight ways in which equity issues should be considered in designing assessments. We also discuss diversity issues in greater detail when we turn to implementation in Chapter 7.

DIMENSIONS OF THE FRAMEWORK

The framework is organized by its three primary dimensions: (1) scientific and engineering practices, (2) crosscutting concepts, and (3) disciplinary core ideas: see Box 2-1. This three-part structure signals an important shift for science education and presents the primary challenge for assessment design: to find a way to capture and support students' developing proficiency along the intertwined dimensions.

BOX 2-1

THE THREE DIMENSIONS OF THE FRAMEWORK

1 Scientific and Engineering Practices

1. Asking questions (for science) and defining problems (for engineering)
2. Developing and using models
3. Planning and carrying out investigations
4. Analyzing and interpreting data
5. Using mathematics and computational thinking
6. Constructing explanations (for science) and designing solutions (for engineering)
7. Engaging in argument from evidence
8. Obtaining, evaluating, and communicating information

2 Crosscutting Concepts

1. *Patterns.* Observed patterns of forms and events guide organization and classification, and they prompt questions about relationships and the factors that influence them.
2. *Cause and effect: Mechanism and explanation.* Events have causes, sometimes simple, sometimes multifaceted. A major activity of science is investigating and explaining causal relationships and the mechanisms by which they are mediated. Such mechanisms can then be tested across given contexts and used to predict and explain events in new contexts.
3. *Scale, proportion, and quantity.* In considering phenomena, it is critical to recognize what is relevant at different measures of size, time, and energy and to recognize how changes in scale, proportion, or quantity affect a system's structure or performance.
4. *Systems and system models.* Defining the system under study—specifying its boundaries and making explicit a model of that system—provides tools for understanding and testing ideas that are applicable *throughout science and engineering.*
5. *Energy and matter: Flows, cycles, and conservation.* Tracking fluxes of energy and matter into, out of, and within systems helps one understand the systems' possibilities and limitations.

Dimension 1: Scientific and Engineering Practices

Dimension 1 identifies eight important practices used by scientists and engineers, such as modeling, developing explanations or solutions, and engaging in argumentation. The framework emphasizes that students need to actively engage in these scientific and engineering practices in order to truly understand the core ideas in the disciplines. The introduction of practices is not a rejection of the importance

6. *Structure and function.* The way in which an object or living thing is shaped and its substructure determine many of its properties and functions.
7. *Stability and change.* For natural and built systems alike, conditions of stability and determinants of rates of change or evolution of a system are critical elements of study.

3   Disciplinary Core Ideas

Physical Sciences
PS1: Matter and its interactions
PS2: Motion and stability: Forces and interactions
PS3: Energy
PS4: Waves and their applications in technologies for information transfer

Life Sciences
LS1: From molecules to organisms: Structures and processes
LS2: Ecosystems: Interactions, energy, and dynamics
LS3: Heredity: Inheritance and variation of traits
LS4: Biological evolution: Unity and diversity

Earth and Space Sciences
ESS1: Earth's place in the universe
ESS2: Earth's systems
ESS3: Earth and human activity

Engineering, Technology, and Applications of Science
ETS1: Engineering design
ETS2: Links among engineering, technology, science, and society

SOURCE: National Research Council (2012a, pp. 3, 84).

of engaging students in inquiry as a component of science learning but rather a clarification that highlights the diversity of what scientists actually do.

The framework asserts that students cannot appreciate the nature of scientific knowledge without directly experiencing and reflecting on the practices that scientists use to investigate and build models and theories about the world. Nor can they appreciate the nature of engineering unless they engage in the practices that engineers use to design and build systems. The opportunity to learn by experiencing and reflecting on these practices, the framework's authors note is important because it helps students understand that science and engineering are not a matter of applying rote procedures. Engaging in and reflecting on the practices will help students see science as an iterative process of empirical investigation, evaluation of findings, and the development of explanations and solutions. Likewise, it will help students see engineering—a process of developing and improving a solution to a design problem—as both creative and iterative.

Dimension 2: Crosscutting Concepts

The framework identifies seven crosscutting concepts that can help students link knowledge from the various disciplines as they gradually develop a coherent and scientific view of the world. These crosscutting concepts are fundamental to understanding science and engineering, but they have rarely been taught or have not been taught in a way that fosters understanding of their cross-disciplinary utility and importance. Explicit attention to these concepts can help students develop an organizational framework for connecting knowledge across disciplines and developing integrated understanding of what they learn in different settings. The crosscutting concepts will be reinforced when they are addressed in the context of many different disciplinary core ideas. The framework posits that if this is done intentionally, using consistent language across years of schooling, students can come to recognize how the concepts apply in different contexts and begin to use them as tools to examine new problems. The idea that crosscutting concepts are fundamental to understanding science and engineering is not a new idea. Chapter 11 of *Science for All Americans* could not be clearer about the importance of crosscutting concepts and how they apply across the different areas of science.[2]

---

[2]See http://www.project2061.org/publications/sfaa/online/chap11.htm [March 2014].

Developing Assessments for the Next Generation Science Standards

Dimension 3: Disciplinary Core Ideas

The framework identifies disciplinary core ideas for the physical, life, and earth and space sciences and for engineering, technology, and applications of science. The framework makes clear that the purpose of science education is not to teach all the details—an impossible task—but to prepare students with sufficient core knowledge and abilities so that they can acquire and evaluate additional information on their own or as they continue their education.

The dimension of core ideas is extremely important. Education structured around a limited number of core ideas allows the time necessary for students to explore ideas in greater depth at each grade level and engage in the full range of practices. This dimension is in part a practical idea that has gained currency as people have recognized that curricula and standards that cover many details are too broad to provide guidance about priorities and can lead to instruction that is "a mile wide and an inch deep" (Schmidt et al., 1999). Research on science learning also supports the idea that learning should be linked to organizing structures (National Research Council, 2007).

INTEGRATION: THREE-DIMENSIONAL SCIENCE LEARNING

The framework emphasizes that science and engineering education should support the integration of disciplinary core ideas and crosscutting concepts with the practices needed to engage in scientific inquiry and engineering design.[3] In this report, we refer to this integration of content knowledge, crosscutting concepts, and practices as "three-dimensional science learning," or more simply "three-dimensional learning." That is, during instruction, students' engagement in the practices should always occur in the context of a core idea and, when possible, should also connect to crosscutting concepts. Both practices and crosscutting ideas are viewed as tools for addressing new problems as well as topics for learning in themselves. Students need to experience the use of these tools in multiple contexts in order to develop the capacity to wield them flexibly and effectively in new problem contexts—an important goal of science learning (National Research Council, 2000, 2007).

To support this kind of science learning, standards, curriculum materials, instruction, and assessments have to integrate all three dimensions. The frame-

---

[3]We note that students cannot engage in all the practices of science and engineering in the ways that scientists and engineers carry them out. Thus, the practices we refer to in this report are approximations of the practices through which scientists and engineers generate and revise their understandings of natural and designed systems.

work thus recommends that standards take the form of performance expectations that specify what students should know and be able to do in terms that clearly blend or coordinate practices with disciplinary core ideas and crosscutting concepts.[4] Assessment tasks, in turn, have to be designed to provide evidence of students' ability to use the practices, to apply their understanding of the crosscutting concepts, and to draw on their understanding of specific disciplinary ideas, all in the context of addressing specific problems.

In developing the NGSS, development teams from 26 states and the consultants coordinated by Achieve, Inc., elaborated the framework's guidelines into a set of performance expectations that include descriptions of the ways in which students at each grade are expected to use both the practices and crosscutting concepts combined with the knowledge they are expected to have of the core ideas. The performance expectations are available in two organizational arrangements, by disciplinary core idea or by topic. Each presents related ideas in such a way that it is possible to read through clusters of performance expectations related to, for example, a particular aspect of a disciplinary core idea at each grade or grade band. Each performance expectation asks students to use a specific practice, and perhaps also a crosscutting concept, in the context of a disciplinary core idea. Across the set of expectations for a given grade level, each practice and each crosscutting idea appears in multiple standards.

To illustrate, Box 2-2 shows performance expectations for 2nd-grade students related to matter and its interactions. The top section (considered the assessable component) lists four performance expectations that describe what 2nd-grade students who demonstrate the desired grade-level understanding in this area know and can do. The three vertical columns below and in the center (called "foundation boxes") provide the connections to the three dimensions, listing the specific practices students would use and the relevant specific core ideas and crosscutting concepts for this grade level. The text in these boxes expands and explains the performance expectations in terms of each of the three framework dimensions.[5]

The framework argues that disciplinary core ideas should be systematically revisited in new contexts across time to allow students to apply, extend, and develop more sophisticated understanding of them. Instruction should thus care-

---

[4]The performance expectations recommended in the framework are based on the model put forward in *Science: College Board Standards for College Success* (College Board, 2009).

[5]The NGSS also show the connections to performance expectation for other core ideas for the 2nd grade and to related performance expectations for later grade levels, as well as links to elements of the Common Core State Standards in English language arts and mathematics.

fully build ideas across years and between science disciplines. Instead of treating a large number of independent topics, instruction should guide students along pathways through learning progressions. This approach calls for standards, curriculum materials, and assessments that are coherent across time so that they can both help students build increasingly sophisticated understandings of the core ideas across multiple grades and support students in making connections among core ideas in different disciplines.

Learning Progressions: Developing Proficiency Over Time

Research on learning shows that to develop a coherent understanding of scientific explanations of the world, students need sustained opportunities to engage in the practices, work with the underlying ideas, and appreciate the interconnections among these practices and ideas over a period of years, not weeks or months (National Research Council, 2007). Researchers and science educators have applied this insight into how students learn in descriptions of the way understanding of particular content matures over time, called *learning progressions*. Learning progressions may provide the basis for guidance on the instructional supports and experiences needed for students to make progress (as argued in Gotwals and Songer, 2013; Corcoran et al., 2009; National Research Council, 2007; Smith et al., 2006).

Learning progressions are anchored at one end by what is known about the concepts and reasoning students have as they enter school. At the other end, learning progressions are anchored by societal expectations about what students should understand about science by the end of high school. Learning progressions describe the developing understandings that students need as they progress between these anchor points—the ideas and practices that contribute to building a more mature understanding. They often also address common misunderstandings and describe a continuum of increasing degrees of conceptual sophistication that are common as students if they are exposed to suitable instruction (National Research Council, 2007).

The framework builds on this idea by specifying grade-band endpoint targets at grades 2, 5, 8, and 12 for each component of each core idea. The grade-band endpoints are based on research and on the framework committee's judgments about grade appropriateness. Most of the progressions described in the NGSS (which are based on the endpoints described in the framework) were not primarily based on empirical research about student learning of specific material because such research is available only for a limited number of topics (see

BOX 2-2

EXAMPLE OF A PERFORMANCE EXPECTATION IN THE NGSS: MATTER AND ITS INTERACTIONS FOR STUDENTS IN 2ND GRADE

2-PS1 Matter and Its Interactions

*The performance expectations marked with an asterisk integrate traditional science content with engineering through a Practice or Disciplinary Core Idea. The section titled "Disciplinary Core Ideas" is reproduced verbatim from *A Framework for K-12 Science Education: Practices, Crosscutting Concepts, and Core Ideas*. Integrated and reprinted with permission from the National Academy of Sciences.

Students who demonstrate understanding can:

2-PS1-1. Plan and conduct an investigation to describe and classify different kinds of materials by their observable properties. [Clarification Statement: Observations could include color, texture, hardness, and flexibility. Patterns could include the similar properties that different materials share.]

2-PS1-2. Analyze data obtained from testing different materials to determine which materials have the properties that are best suited for an intended purpose.* [Clarification Statement: Examples of properties could include strength, flexibility, hardness, texture, and absorbency.] [Assessment Boundary: Assessment of quantitative measurements is limited to length.]

2-PS1-3. Make observations to construct an evidence-based account of how an object made of a small set of pieces can be disassembled and made into a new object. [Clarification Statement: Examples of pieces could include blocks, building bricks, or other assorted small objects.]

2-PS1-4. Construct an argument with evidence that some changes caused by heating or cooling can be reversed and some cannot. [Clarification Statement: Examples of reversible changes could include materials such as water and butter at different temperatures. Examples of irreversible changes could include cooking an egg, freezing a plant leaf, and heating paper.]

Science and Engineering Practices

Planning and Carrying Out Investigations

Planning and carrying out investigations to answer questions or test solutions to problems in K-2 builds on prior experiences and progresses to simple investigations, based on fair tests, which provide data to support explanations or design solutions.

- Plan and conduct an investigation collaboratively to produce data to serve as the basis for evidence to answer a question. (2-PS1-1)

Analyzing and Interpreting Data

Analyzing data in K-2 builds on prior experiences and progresses to collecting, recording, and sharing observations.

- Analyze data from tests of an object or tool to determine if it works as intended. (2-PS1-2)

Constructing Explanations and Designing Solutions

Constructing explanations and designing solutions in K-2 builds on prior experiences and progresses to the use of evidence and ideas in constructing evidence-based accounts of natural phenomena and designing solutions.

- Make observations (firsthand or from media) to construct an evidence-based account for natural phenomena. (2-PS1-3)

Engaging in Argument from Evidence

Engaging in argument from evidence in K-2 builds on prior experiences and progresses to comparing ideas and representations about the natural and designed world(s).

- Construct an argument with evidence to support a claim. (2-PS1-4)

Disciplinary Core Ideas

PS1.A: Structure and Properties of Matter

- Different kinds of matter exist and many of them can be either solid or liquid, depending on temperature. Matter can be described and classified by its observable properties. (2-PS1-1)
- Different properties are suited to different purposes. (2-PS1-2, 2-PS1-3)
- A great variety of objects can be built up from a small set of pieces. (2-PS1-3)

PS1.B: Chemical Reactions

- Heating or cooling a substance may cause changes that can be observed. Sometimes these changes are reversible, and sometimes they are not. (2-PS1-4)

Crosscutting Concepts

Patterns

- Patterns in the natural and human-designed world can be observed. (2-PS1-1)

Cause and Effect

- Events have causes that generate observable patterns. (2-PS1-4)
- Simple tests can be designed to gather evidence to support or refute student ideas about causes. (2-PS1-2)

Energy and Matter

- Objects may break into smaller pieces and be put together into larger pieces, or change shapes. (2-PS1-3)

- - - - - - - - - - - - - - -

Connections to Engineering, Technology, and Applications of Science

Influence of Engineering, Technology, and Science on Society and the Natural World

- Every human-made product is designed by applying some knowledge of the natural world and is built using materials derived from the natural world. (2-PS1-2)

SOURCE: NGSS Lead States (2013). Copyright 2013 Achieve, Inc. All rights reserved. Available: http://www.nextgenscience.org/2ps1-matter-interactions [March 2014].

Corcoran et al., 2009).[6] Thus, the framework and the NGSS drew on available research, as well as on experience from practice and other research- and practice-based documents (American Association for the Advancement of Science, 2001, 2007; National Research Council, 1996). The NGSS endpoints provide a set of initial hypotheses about the progression of learning for the disciplinary core ideas (National Research Council, 2012a, p. 33). An example, for ideas about how energy for life is derived from food, is shown in Box 2-3.

For the practices and crosscutting concepts, the framework provides sketches of possible progressions for learning each practice or concept, but it does not indicate the expectations at any particular grade level. The NGSS built on those sketches and provide a matrix that defines what each practice might encompass at each grade level, as well as a matrix that defines the expected uses of each

---

[6]The American Association for the Advancement of Science (2001, 2007) is another source of progressions of learning that are based on available research supplemented with expert judgment.

crosscutting concept for students at each grade level through 5th grade and in grade bands for middle school and high school.

The progressions in the NGSS are not learning progressions as defined in science education research because they neither articulate the instructional support that would be needed to help students achieve them nor provide a detailed description of students' developing understanding. (They also do not identify specific assessment targets, as assessment-linked learning progressions do.) However, they are based on the perspective that instruction and assessments must be designed to support and monitor students as they develop increasing sophistication in their ability to use practices, apply crosscutting concepts, and understand core ideas as they progress across the grade levels.

Assessment developers will need to draw on the idea of developing understanding as they structure tasks for different levels and purposes and build this idea into the scoring rubrics for the tasks. The target knowledge at a given grade level may well involve an incomplete or intermediate understanding of the topic or practice. Targeted intermediate understandings can help students build toward a more scientific understanding of a topic or practice, but they may not themselves be fully complete or correct. They are acceptable stepping stones on the pathways students travel between naïve conceptions and scientists' best current understandings.

Supporting Connections Across Disciplines

A second aspect of coherence in science education lies in the connections among the disciplinary core ideas, such as using understandings about chemical interactions from physical science to explain phenomena in biological contexts. The framework was designed so that when students are working on a particular idea in one discipline, they will already have had experience with the necessary foundational ideas in other disciplines. So, for example, if students are learning about how food is used by organisms in the context of the life sciences in the middle grades, they should already have learned the relevant ideas about chemical transformations in the context of the physical sciences. These connections between ideas in different disciplines are called out in the foundation boxes of the NGSS, which list connections to other disciplinary core ideas at the same grade level, as well as ideas at other grade levels (see Box 2-2, above).

EXAMPLE 1: WHAT IS GOING ON INSIDE ME?

This example of an assessment task illustrates the concept of three-dimensional science learning, the kinds of instructional experiences that are needed to support its development, and the assessment tasks that can provide documentation of this kind of learning.[7] It also shows how a performance expectation can be used to develop an assessment task and the associated scoring rubric. Specifically, it illustrates how students' classroom investigations yield products that can be used as formative assessments of their understanding of and ability to connect disciplinary core ideas.

Instructional Context

The curriculum materials for the 7th-grade unit, "What Is Going on Inside Me," were developed as part of the 3-year middle school curriculum series developed by the Investigating and Questioning our World through Science and Technology (IQWST) project (Krajcik et al., 2008b; Shwartz et al., 2008). IQWST units were designed to involve middle school students in investigation, argumentation, and model building as they explore disciplinary core ideas in depth. IQWST units begin with a driving question, and students investigate phenomena and engage in scientific argumentation to develop explanations through class consensus. In this 7th-grade unit on the human body (Krajcik et al., 2013), the students are on a hunt through the body to find out where the food is going and how the body gets matter and the energy out of that food. Along the way, they also discover that oxygen is required for the production of energy from food.

When students in the middle grades study how food is used, they have to draw on ideas from physical science, such as conservation of matter, transformation of energy, and chemical reactions, if they are to develop the explanatory core idea in the framework. Understanding how energy and matter cycle and flow is a tool for understanding the functioning of any system—so these are crosscutting concepts as well. In this example, the target for learning is not just the idea that humans—like other animals—use food to provide energy, but also a reasoned explanation that the release of this energy must involve a chemical reaction,

---

[7]As noted in Chapter 1, we use examples of assessment tasks to illustrate the discussion. This is the first of the seven examples, which are numbered consecutively across Chapters 2, 3, and 4. Like all of our examples, this one is drawn from work done before the framework and the NGSS were available, but the expectations that drove its design are very similar to those in the framework and the NGSS.

and an evidence-based argument for this explanatory account. This explanation requires building knowledge that connects core ideas across several disciplines, from physical sciences to life sciences, as tools to develop and defend the explanation with an argument based on evidence.

In this 8-week investigation, the teacher introduces a general question about what happens inside the body that helps humans do the things they do. The curriculum materials guide students to link this question to their real-world experiences, observations, and activities. Students are expected to develop an explanation for where in the body energy and building materials are obtained from food and how this happens as they progress through all of the activities in the unit.

Teachers support the students through a series of investigations in which pursuing the driving question leads to more specific questions, motivating particular investigations focused on cell growth, what cells need to survive, identifying what materials can get into and out of a cell, and so on. Thus, each step involves questions that teachers develop with their students. Each step helps students incrementally build and extend their model and explanation of the central phenomena as they answer the driving question (Krajcik et al., 2008). Together, they incrementally build evidence and an argument for the explanation that food is broken down and transported through the body to all the cells, where a chemical reaction occurs that uses oxygen and glucose to release energy for use by the cells.

Thus, the question is broadened to also track where the oxygen goes and how it is used, as students notice that increased activity in the body is associated with increased oxygen intake. Tracing of the glucose and the oxygen leads to the conclusion that the food and oxygen are going to all the cells of the body and that is where the energy is released. Teachers support students in figuring out that the only thing that could rearrange the matter in the ways needed and release the energy that the cells appear to be using to do their work is through a chemical reaction. Assembling these arguments depends critically on understandings about energy and chemical reactions that they have developed earlier: see Table 2-1.

Assessment

The assessment portion of the example focuses not only on the important claims students have identified (e.g., that oxygen is used by cells) but also on students' proficiency with providing an argument for an explanatory mechanism that connects relevant scientific ideas from different disciplines (e.g., a chemical reaction is needed to release stored energy from food, and oxygen is a component of that

TABLE 2-1 Drawing on Prior Principles from Life and Physical Sciences to Construct a Biological Argument That Supports an Explanation for Where and How Oxygen Is Used in the Body

Component of Core Idea	NGSS DCI	How the Idea Is Used in the Argument
Food provides living things with building materials and energy.	LS1.C (grade 5)	Something must be going on in the body that uses food, and somehow gets the matter to be used in growth, and the energy to be used for all body functions.
All matter is made of particles; matter cannot be created or destroyed.	PS1.A (grade 5)	The increased mass in growth must come from somewhere, so it must be from the food input to the body.
Energy cannot be created or destroyed, but can be transferred from one part of a system to another, and converted from one form to another.	PS3.B (grade 8)	The only way for the body to get energy is to get it from somewhere else, either transfer or conversion of energy.
Chemical reactions can rearrange matter into different combinations, changing its properties.	PS3.B (grade 8)	To use the mass in food, a chemical reaction must be taking place to rearrange the substances.
Chemical reaction can convert energy from stored energy to other forms of energy.	PS1.B, PS3.A (grade 8)	There must be a chemical reaction going on to get the stored energy in the food into a form usable by the body.
One type of chemical reaction that can convert stored energy to other forms is when some substances combine with oxygen in burning.	PS3.D (grade 8)	The oxygen that is shipped around the body along with the broken-down food must be being used in a chemical reaction to convert the stored energy in the food molecules.

NOTE: LS = life sciences, NGSS DCI = Next Generation Science Standards, Disciplinary Core Ideas, and PS = physical sciences.
SOURCE: Adapted from Krajcik et al. (2013), National Research Council (2012a), and NGSS Lead States (2013).

chemical reaction). In other words, the assessments (described below) are designed to assess three-dimensional learning.

In national field trials of IQWST, 7th- and 8th-grade students were given an assessment task, which was embedded in the curriculum that reflected the performance expectation shown in Box 2-4. When this assessment was given, students had established that food is broken down into glucose and other components and that the circulatory system distributes glucose so that it reaches each cell in the body. Students' experiments with osmosis had enabled them to conclude that both water and glucose could enter the cell, and experiments with yeast (as a model system for human cells) had led students to establish that cells could use the

PERFORMANCE EXPECTATION FOR UNDERSTANDING OXYGEN USE IN THE BODY

Performance Expectation: Construct and argue for an explanation for why animals breathe out less oxygen than the air they breathe in.

Science and Engineering Practices
- Constructing Explanations and Designing Solutions: Construct explanations and design solutions supported by multiple sources of evidence consistent with scientific knowledge, principles, and theories.
- Engaging in Argument from Evidence: Construct a convincing argument that supports or refutes claims for explanations or solutions about the natural and designed world. Use oral and written arguments supported by empirical evidence and reasoning to support or refute.

Crosscutting Concepts: Energy and Matter

- Matter is conserved because atoms are conserved in physical and chemical processes. Within a natural or designed system, the transfer of energy drives the motion and/or cycling of matter.
- Energy may take different forms (e.g., energy in fields, thermal energy, energy of motion). The transfer of energy can be tracked as energy flows through a designed or natural system.

Disciplinary Core Ideas: LS1.C: Organization for Matter and Energy Flow in Organisms

- Within individual organisms, food moves through a series of chemical reactions in which it is broken down and rearranged to form new molecules, to support growth or to release energy.
- In most animals and plants, oxygen reacts with carbon-containing molecules (sugars) to provide energy and produce carbon dioxide; anaerobic bacteria achieve their energy needs in other chemical processes that do not need oxygen.

SOURCES: Adapted from Krajcik et al. (2013) and National Research Council (2012a).

glucose for energy and growth, and that this process released waste in the form of carbon dioxide gas. Students had also established that increased energy needs (such as physical activity) are associated with increased consumption of air, and that exhaled air contains proportionally less oxygen than the air in the room.

Students were then asked to synthesize their findings in a written argument in response to the following task (Krajcik et al., 2008b):

Solving the mystery: Inspector Bio wants to know what you have figured out about the oxygen that is missing from the air you exhale. Explain to her where the oxygen goes, what uses it, and why. Write a scientific explanation with a claim, sufficient evidence, and reasoning.

Throughout the IQWST curriculum, students learn to write and argue for scientific explanations with a claim, evidence, and reasoning—that is, to incorporate both the construction of an explanation and presentation of an argument for that explanation in their responses (see Berland and Reiser, 2009; McNeill and Krajcik, 2008; Krajcik et al., 2013). Below is a typical response from an 8th-grade student (collected during IQWST field trials) that demonstrates application of the physical science ideas of both energy and matter to explain the oxygen question.

After being inhaled, oxygen goes through the respiratory system, then the circulation system or blood, and goes throughout the body to all the cells. Oxygen is used to burn the food the body needs and get energy for the cells for the body to use. For anything to burn, it must have energy and oxygen. To then get the potential energy in food, the body needs oxygen, because it is a reactant. When we burned the cashew, the water above it increased, giving it thermal energy and heating it up. Therefore, food is burned with oxygen to get energy.

This response shows both what the student currently understands and that he or she drew on evidence from the activity of burning a cashew and thereby heating water. It also illustrates the sort of incomplete target understanding that we have discussed: the student considers the food to contain potential energy but cannot elaborate how the chemical reaction converts the energy to a form cells can use. This conception is acceptable at the middle school level but will need refinement in later grades.

The IQWST materials suggest a scoring rubric for this task: see Box 2-5. The performance expectation and the scoring rubric also show how the assessment measures students' ability to draw on core ideas from multiple disciplines by asking for an argument and explanation about a phenomenon that requires bringing the physical science understanding to bear on an argument in the biological context. This example shows that, with appropriate prior instruction, students can tackle tasks that assess three-dimensional science learning, that is, tasks that ask them to use science practices in the context of crosscutting concepts and disciplinary core ideas. Furthermore, it shows that classroom engagement in practices (in this case, supporting an explanation with argument from evidence) provides products (in this case, written responses to a probe question) that can be used to evaluate student learning.

SCORING RUBRIC (CRITERIA) FOR PERFORMANCE EXPECTATION ON OXYGEN USE IN THE BODY

Level 0: Missing or only generic reasons for survival (e.g., to breathe, for living)

Level 1: Oxygen used to get energy or used with food for energy; no physical science mechanism presented to get energy

Level 2: Oxygen used in a chemical reaction (or "burning") to get energy, but an incomplete description of matter and energy physical science (e.g., "burns the oxygen" without mentioning food or glucose or "react with glucose" but no account of energy)

Level 3: Full account, using physical science ideas including both the matter and energy accounts—oxygen is combined in a chemical reaction with food or glucose that includes a conversion of the stored energy in food to forms usable by the cells

SOURCE: Adapted from Krajcik et al. (2013).

CONCLUSIONS

The framework acknowledges that the new vision for science teaching and learning poses challenges for assessment and will require significant changes to current assessment approaches. The example above is the first of several we use to illustrate the specific changes we believe will be needed; it also illustrates that assessment must be considered as part of the overall system of science education. The framework emphasizes the widely shared understanding that the major components of the science education system (curriculum, instruction, teacher development, and assessment) are tightly linked and interdependent, and it advocates a standards-based system that is coherent horizontally (across classrooms at a given grade level), vertically (across levels of control and aggregation of scores, such as across schools, districts, and a state), and developmentally (across grade levels). The framework also follows an earlier report (National Research Council, 2006) in calling for a coherent system of assessments that combines multiple approaches (e.g., including both large-scale and classroom-based assessments) to meet a range of goals (e.g., formative and summative assessments of student learning, program evaluation) in an integrated and effective way. Given the complexity of the assessment challenge, the framework emphasizes that changes will likely need to be phased in over time.

We offer four conclusions about three specific challenges for design and development of assessments to meet the goals of the framework and the NGSS.

Assessing Three-Dimensional Learning

Assessing three-dimensional learning is perhaps the most significant challenge because it calls for assessment tasks that examine students' performance of a practice at the same time that they are working with disciplinary core ideas and crosscutting concepts. Meeting this challenge can best be accomplished through the use of assessment tasks that comprise multiple related questions, which we refer to as "multicomponent tasks."

CONCLUSION 2-1 Measuring the three-dimensional science learning called for in the framework and the Next Generation Science Standards requires assessment tasks that examine students' performance of scientific and engineering practices in the context of crosscutting concepts and disciplinary core ideas. To adequately cover the three dimensions, assessment tasks will generally need to contain multiple components (e.g., a set of interrelated questions). It may be useful to focus on individual practices, core ideas, or crosscutting concepts in the various components of an assessment task, but, together, the components need to support inferences about students' three-dimensional science learning as described in a given performance expectation.

Assessing the Development of Three-Dimensional Learning Over Time

The framework emphasizes that competence in science develops cumulatively over time and increases in sophistication and power. The framework calls for curricula and instruction that are planned in a coherent way to help students progress along a path toward more sophisticated understanding of core concepts over the course of the entire K-12 grade span. Students' intermediate steps along this path may not reflect accurate scientific understanding, but they will reflect increasingly sophisticated approximations of scientific explanations of phenomena.

Thus, what needs to be assessed is what point a student has reached along a sequence of progressively more complex understandings of a given core idea, and successively more sophisticated applications of practices and crosscutting concepts. This is a relatively unfamiliar idea in the realm of science assessments, which have more often been designed to measure whether students at a given grade level do or do not know particular content (facts). To meet this new goal, assessments will have to reflect both what understanding is expected at a particular grade level and the intermediate understandings that may be appropriate at other levels. This idea of intermediate understanding is particularly important for formative or in-class assessment tools (see Chapter 3).

CONCLUSION 2-2 The Next Generation Science Standards require that assessment tasks be designed so that they can accurately locate students along a sequence of progressively more complex understandings of a core idea and successively more sophisticated applications of practices and crosscutting concepts.

Breadth and Depth of Content

The third challenge is to develop assessment tasks that adequately address all elements of all three dimensions and cover all of the performance expectations for a given grade level. The amount of science knowledge specified in the core ideas alone is demanding. The possible ways the ideas might be combined with the practices and crosscutting concepts into performance expectations even for a single grade would yield an even greater range of possible targets for assessment. Moreover, both competence in using the practices and understanding of core ideas need to develop across the grade levels. The NGSS limit the number of performance expectations by choosing to define particular combinations of practices with aspects of a core idea, but there is still a large amount of material to assess. In addition, the time needed for students to undertake the type of multicomponent tasks that can assess a single performance expectation is much greater than the time for a single multiple-choice item testing a particular piece of knowledge.

CONCLUSION 2-3 The Next Generation Science Standards place significant demands on science learning at every grade level. It will not be feasible to assess all of the performance expectations for a given grade level with any one assessment. Students will need multiple—and varied—assessment opportunities to demonstrate their competence on the performance expectations for a given grade level.

The performance expectations in the NGSS help to narrow the scope of what needs to be assessed, but they are complex in terms of the concepts students need to call on in order to demonstrate mastery. Thus, more than one assessment task may be needed to adequately assess mastery of a given performance expectation, and multiple tasks will be needed to assess the progress of learning all aspects of a particular core idea. We note also that to assess both understanding of core knowledge and facility with a practice, assessments may need to probe students' use of a given practice in more than one disciplinary context. Furthermore, although the practices are described separately, they generally function in concert, such as when students present an argument based on a model and provide some corroborating evidence in support of an explanation, or when students use mathematics as they analyze data. This overlap means that in some cases assessment tasks may need to be designed around a cluster of related performance expectations. Assessment tasks that attempt to test practices in strict isolation from one another may not be meaningful as assessments of the three-dimensional science learning called for by the NGSS.

CONCLUSION 2-4 **Effective evaluation of three-dimensional science learning requires more than a one-to-one mapping between the Next Generation Science Standards (NGSS) performance expectations and assessment tasks. More than one assessment task may be needed to adequately assess students' mastery of some performance expectations, and any given assessment task may assess aspects of more than one performance expectation. In addition, to assess both understanding of core knowledge and facility with a practice, assessments may need to probe students' use of a given practice in more than one disciplinary context. Assessment tasks that attempt to test practices in strict isolation from one another may not be meaningful as assessments of the three-dimensional science learning called for by the NGSS.**

3

ASSESSMENT DESIGN AND VALIDATION

Measuring science content that is integrated with practices, as envisioned in *A Framework for K-12 Science Education: Practices, Crosscutting Concepts, and Core Ideas* (National Research Council, 2012a, hereafter referred to as "the framework") and the *Next Generation Science Standards: For States, By States* (NGSS Lead States, 2013), will require a careful and thoughtful approach to assessment design. This chapter focuses on strategies for designing and implementing assessment tasks that measure the intended practice skills and content understandings laid out in the Next Generation Science Standards (NGSS) performance expectations.

Some of the initial stages of assessment design have taken place as part of the process of writing the NGSS. For example, the NGSS include progressions for the sequence of learning, performance expectations for each of the core ideas addressed at a given grade level or grade band, and a description of assessable aspects of the three dimensions addressed in the set of performance expectations for that topic. The performance expectations, in particular, provide a foundation for the development of assessment tasks that appropriately integrate content and practice. The NGSS performance expectations also usually include boundary statements that identify limits to the level of understanding or context appropriate for a grade level and clarification statements that offer additional detail and examples. But standards and performance expectations, even as explicated in the NGSS, do not provide the kind of detailed information that is needed to create an assessment.

The design of valid and reliable science assessments hinges on multiple elements that include but are not restricted to what is articulated in disciplinary frameworks and standards (National Research Council, 2001; Mislevy and Haertel, 2006). For example, in the design of assessment items and tasks related to the NGSS performance expectations, one needs to consider (1) the kinds of conceptual models and evidence that are expected of students; (2) grade-level-appropriate contexts for assessing the performance expectations; (3) essential and optional task design features (e.g., computer-based simulations, computer-based animations, paper-pencil writing and drawing) for eliciting students' ideas about the performance expectation; and (4) the types of evidence that will reveal levels of students' understandings and skills.

Two prior National Research Council reports have addressed assessment design in depth and offer useful guidance. In this chapter, we draw from *Knowing What Students Know* (National Research Council, 2001) and *Systems for State Science Assessment* (National Research Council, 2006) in laying out an approach to assessment design that makes use of the fundamentals of cognitive research and theory and measurement science. We first discuss assessment as a process of reasoning from evidence and then consider two contemporary approaches to assessment development—evidence-centered design and construct modeling—that we think are most appropriate for designing individual assessment tasks and collections of tasks to evaluate students' competence relative to the NGSS performance expectations.[1] We provide examples of each approach to assessment task design. We close the chapter with a discussion of approaches to validating the inferences that can be drawn from assessments that are the product of what we term a principled design process (discussed below).

ASSESSMENT AS A PROCESS OF EVIDENTIARY REASONING

Assessment specialists have found it useful to describe assessment as a process of reasoning from evidence—of using a representative performance or set of performances to make inferences about a wider set of skills or knowledge. The process of collecting evidence to support inferences about what students know and can do

---

[1]The word "construct" is generally used to refer to concepts or ideas that cannot be directly observed, such as "liberty." In the context of educational measurement, it is used more specifically to refer to a particular body of content (knowledge, understanding, or skills) that an assessment is designed to measure. It can be used to refer to a very specific aspect of tested content (e.g., the water cycle) or a much broader area (e.g., mathematics).

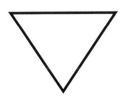

Observation

Interpretation

Cognition

FIGURE 3-1 The three elements involved in conceptualizing assessment as a process of reasoning from evidence.
SOURCE: Adapted from National Research Council (2001, p. 44).

is fundamental to all assessments—from classroom quizzes, standardized achievement tests, or computerized tutoring programs, to the conversations students have with their teachers as they work through an experiment. The Committee on the Cognitive Foundations of Assessment (National Research Council, 2001) portrayed this process of reasoning from evidence in the form of an assessment triangle: see Figure 3-1.

The triangle rests on cognition, defined as a "theory or set of beliefs about how students represent knowledge and develop competence in a subject domain" (National Research Council, 2001, p. 44). In other words, the design of the assessment should begin with specific understanding not only of which knowledge and skills are to be assessed, but also of how understanding and competence develop in the domain of interest. For the NGSS, the cognition to be assessed consists of the the practices, the crosscutting concepts, and disciplinary core ideas as they are integrated in the performance expectations.

A second corner of the triangle is observation of students' capabilities in the context of specific tasks designed to show what they know and can do. The capabilities must be defined because the design and selection of the tasks need to be tightly linked to the specific inferences about student learning that the assessment is intended to support. It is important to emphasize that although there are various factors that assessments could address, task design should be based on an explicit definition of the precise aspects of cognition the assessment is targeting. For example, assessment tasks that engage students in applying the three-dimensional learning (described in Chapter 2) could possibly yield information about how students use or apply specific practices, crosscutting concepts, disciplinary core ideas, or combinations of these. If the intended constructs are clearly specified, the design of a specific task and its scoring rubric can support clear inferences about students' capabilities.

The third corner of the triangle is interpretation, meaning the methods and tools used to reason from the observations that have been collected. The method used for a large-scale standardized test might involve a statistical model. For a classroom assessment, it could be a less formal method of drawing conclusions about a student's understanding on the basis of the teacher's experiences with the student, or it could provide an interpretive framework to help make sense of different patterns in a student's contributions to practice and responses to questions.

The three elements are presented in the form of a triangle to emphasize that they are interrelated. In the context of any assessment, each must make sense in terms of the other two for the assessment to produce sound and meaningful results. For example, the questions that shape the nature of the tasks students are asked to perform should emerge logically from a model of the ways learning and understanding develop in the domain being assessed. Interpretation of the evidence produced should, in turn, supply insights into students' progress that match up with that same model. Thus, designing an assessment is a process in which every decision should be considered in light of each of these three elements.

Construct-Centered Approaches to Assessment Design

Although it is very valuable to conceptualize assessment as a process of reasoning from evidence, the design of an actual assessment is a challenging endeavor that needs to be guided not only by theory and research about cognition, but also by practical prescriptions regarding the processes that lead to a productive and potentially valid assessment for a particular use. As in any design activity, scientific knowledge provides direction and constrains the set of possibilities, but it does not prescribe the exact nature of the design, nor does it preclude ingenuity in achieving a final product. Design is always a complex process that applies theory and research to achieve near-optimal solutions under a series of multiple constraints, some of which are outside the realm of science. For educational assessments, the design is influenced in important ways by such variables as purpose (e.g., to assist learning, to measure individual attainment, or to evaluate a program), the context in which it will be used (for a classroom or on a large scale), and practical constraints (e.g., resources and time).

The tendency in assessment design has been to work from a somewhat "loose" description of what it is that students are supposed to know and be able to do (e.g., standards or a curriculum framework) to the development of tasks or problems for them to answer. Given the complexities of the assessment design process, it is unlikely that such a process can lead to a quality assessment without a great deal of artistry, luck, and trial and error. As a consequence, many assessments fail to adequately represent the cognitive constructs and content to be covered and leave room for considerable ambiguity about the scope of the inferences that can be drawn from task performance. If it is recognized that assessment is an evidentiary reasoning process, then a more systematic process of assessment design can be used. The assessment triangle provides a conceptual mapping of the nature of assessment, but it needs elaboration to be useful for constructing assessment tasks and assembling them into tests. Two groups of researchers have generated frameworks for developing assessments that take into account the logic embedded in the assessment triangle. The evidence-centered design approach has been developed by Mislevy and colleagues (see, e.g., Almond et al., 2002; Mislevy, 2007; Mislevy et al., 2002; Steinberg et al., 2003), and the construct-modeling approach has been developed by Wilson and his colleagues (see, e.g., Wilson, 2005). Both use a construct-centered approach to task development, and both closely follow the evidentiary reasoning logic spelled out by the NRC assessment triangle.

A construct-centered approach differs from more traditional approaches to assessment, which may focus primarily on surface features of tasks, such as how they are presented to students, or the format in which students are asked to respond.[2] For instance, multiple-choice items are often considered to be useful only for assessing low-level processes, such as recall of facts, while performance tasks may be viewed as the best way to elicit more complex cognitive processes. However, multiple-choice questions can in fact be designed to tap complex cognitive processes (Wilson, 2009; Briggs et al., 2006). Likewise, performance tasks, which are usually intended to assess higher-level cognitive processes, may inadvertently tap only low-level ones (Baxter and Glaser, 1998; Hamilton et al., 1997; Linn et al., 1991). There are, of course, limitations to the range of constructs that multiple-choice items can assess.

As we noted in Chapter 2, assessment tasks that comprise multiple interrelated questions, or components, will be needed to assess the NGSS performance expectations. Further, a range of item formats, including construct-response and performance tasks, will be essential for the assessment of three-dimensional learning consonant with the framework and the NGSS. A construct-centered approach

---

[2]Messick (1994) distinguishes between task-centered performance assessment, which begins with a specific activity that may be valued in its own right (e.g., an artistic performance) or from which one can score particular knowledge or skills, and construct-centered performance assessment, which begins with a particular construct or competency to be measured and creates a task in which it can be revealed.

focuses on "the knowledge, skills, or other attributes to be assessed" and considers "what behaviors or performances should reveal those constructs and what tasks or situations should elicit those behaviors" (Messick, 1994, p. 16). In a construct-centered approach, the selection and development of assessment tasks, as well as the scoring rubrics and criteria, are guided by the construct to be assessed and the best ways of eliciting evidence about a student's proficiency with that construct.

Both evidence-centered design and construct-modeling approach the process of assessment design and development by:

- analyzing the cognitive domain that is the target of an assessment;
- specifying the constructs to be assessed in language detailed enough to guide task design;
- identifying the inferences that the assessment should support;
- laying out the type of evidence needed to support those inferences;
- designing tasks to collect that evidence, modeling how the evidence can be assembled and used to reach valid conclusions; and
- iterating through the above stages to refine the process, especially as new evidence becomes available.

Both methods are called "principled" approaches to assessment design in that they provide a methodical and systematic approach to designing assessment tasks that elicit student performances that reveal their proficiency. Observation of these performances can support inferences about the constructs being measured. Both are approaches that we judged to be useful for developing assessment tasks that effectively measure content intertwined with practices.

Evidence-Centered Design

The evidence-centered design approach to assessment development is the product of conceptual and practical work pursued by Mislevy and his colleagues (see, e.g., Almond et al., 2002; Mislevy, 2007; Mislevy and Haertel, 2006; Mislevy et al., 2002; Steinberg et al., 2003). In this approach, designers construct an assessment argument that is a claim about student learning that is supported by evidence relevant to the intended use of the assessment (Huff et al., 2010). The claim should be supported by observable and defensible evidence.

Figure 3-2 shows these three essential components of the overall process. The process starts with defining as precisely as possible the claims that one wants

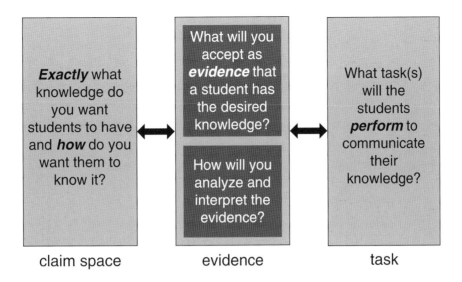

claim space   evidence   task

FIGURE 3-2 Simplified representation of three critical components of the evidence-centered design process and their reciprocal relationships.
SOURCE: Pellegrino et al. (2014, fig. 29.2, p. 576). Reprinted with the permission of Cambridge University Press.

to be able to make about students' knowledge and the ways in which students are supposed to know and understand some particular aspect of a content domain. Examples might include aspects of force and motion or heat and temperature. The most critical aspect of defining the claims one wants to make for purposes of assessment is to be as precise as possible about the elements that matter and to express them in the form of verbs of cognition (e.g., compare, describe, analyze, compute, elaborate, explain, predict, justify) that are much more precise and less vague than high-level cognitive superordinate verbs, such as know and understand. Guiding this process of specifying the claims is theory and research on the nature of domain-specific knowing and learning.

Although the claims one wishes to make or verify are about the student, they are linked to the forms of evidence that would provide support for those claims—the warrants in support of each claim. The evidence statements associated with given sets of claims capture the features of work products or performances that would give substance to the claims. This evidence includes which features need to be present and how they are weighted in any evidentiary scheme (i.e., what matters most and what matters least or not at all). For example, if the evidence in support of a claim about a student's knowledge of the laws of motion is that the student can analyze a physical situation in terms of the forces acting on all the bodies, then the evidence might be a diagram of bodies that is drawn with all the forces labeled, including their magnitudes and directions.

The value of the precision that comes from elaborating the claims and evidence statements associated with a domain of knowledge and skill is clear when one turns to the design of the tasks or situations that can provide the requisite evidence. In essence, tasks are not designed or selected until it is clear what forms of evidence are needed to support the range of claims associated with a given assessment situation. The tasks need to provide all the necessary evidence, and they should allow students to "show what they know" in a way that is as unambiguous as possible with respect to what the task performance implies about their knowledge and skill (i.e., the inferences about students' cognition that are permissible and sustainable from a given set of assessment tasks or items).[3]

As noted above, the NGSS have begun the work of defining such claims about student proficiency by developing performance expectations, but it is only a beginning. The next steps are to determine the observations—the forms of evidence in student work—that are needed to support the claims and then to develop the tasks or situations that will elicit the required evidence. This approach goes beyond the typical approach to assessment development, which generally involves simply listing specific content and skills to be covered and asking task developers to produce tasks related to these topics. The evidence-centered design approach looks at the interaction between content and skills to discern, for example, how students reason about a particular content area or construct. Thus, ideally, this approach yields test scores that are very easy to understand because the evidentiary argument is based not on a general claim that the student "knows the content," but on a comprehensive set of claims that indicate specifically what the student can do within the domain. The claims that are developed through this approach can be guided by the purpose for assessment (e.g., to evaluate a students' progress during a unit of instruction, to evaluate a students' level of achievement at the end of a course) and targeted to a particular audience (e.g., students, teachers).

---

[3]For more information on this approach, see National Research Council (2003), as well as the references cited above.

Evidence-centered design rests on the understanding that the context and purpose for an educational assessment affects the way students manifest the knowledge and skills to be measured, the conditions under which observations will be made, and the nature of the evidence that will be gathered to support the intended inference. Thus, good assessment tasks cannot be developed in isolation; they must be designed around the intended inferences, the observations, the performances that are needed to support those inferences, the situations that will elicit those performances, and a chain of reasoning that will connect them.

Construct Modeling

Wilson (2005) proposes another approach to assessment development: construct modeling. This approach uses four building blocks to create assessments and has been used for assessments of both science content (Briggs et al., 2006; Claesgens et al., 2009; Wilson and Sloane, 2000) and science practices (Brown et al., 2010), as well as to design and test models of the typical progression of understanding of particular concepts (Black et al., 2011; Wilson, 2009). The building blocks are viewed as a guide to the assessment design process, rather than as step-by-step instructions.

The first building block is specification of the construct, in the form of a construct map. Construct maps consist of working definitions of what is to be measured, arranged in terms of consecutive levels of understanding or complexity.[4] The second building block is item design, a description of the possible forms of items and tasks that will be used to elicit evidence about students' knowledge and understanding as embodied in the constructs. The third building block is the outcome space, a description of the qualitatively different levels of responses to items and tasks that are associated with different levels of the construct. The last building block is the measurement model, the basis on which assessors and users associate scores earned on items and tasks with particular levels of the construct; the measurement model relates the scored responses to the constructs. These building blocks are described in a linear fashion, but they are intended to work as elements of a development cycle, with successive iterations producing better coherence among the blocks.[5]

---

[4]When the construct is multidimensional, multiple constructs will be developed, one for each outcome dimension.

[5]For more information on construct modeling, see National Research Council (2003, pp. 89-104), as well as the references cited above.

In the next section, we illustrate the steps one would take in using the two construct-centered approaches to the development of assessment tasks. We first illustrate the evidence-centered design approach using an example developed by researchers at SRI International. We then illustrate the construct-modeling approach using an example from the Berkeley Evaluation and Assessment Research (BEAR) System.

ILLUSTRATIONS OF TASK-DESIGN APPROACHES

In this section, we present illustrations of how evidence-centered design and construct modeling can be used to develop an assessment task. The first example is for students at the middle school level; the second is for elementary school students. In each case, we first describe the underlying design process and then the task.

Evidence-Centered Design—Example 2: Pinball Car Task

Our example of applying evidence-centered design is drawn from work by a group of researchers at SRI International.[6] The task is intended for middle school students and was designed to assess student's knowledge of both science content and practices. The content being assessed is knowledge of forms of energy in the physical sciences, specifically knowledge of potential and kinetic energy and knowledge that objects in motion possess kinetic energy. In the assessment task, students observe the compression of a spring attached to a plunger, the kind of mechanism used to put a ball "in play" in a pinball machine. A student observes that when the plunger is released, it pushes a toy car forward on a racing track. The potential energy in the compressed spring is transformed, on the release of the plunger, into kinetic energy that moves the toy car along the racing track. The student is then asked to plan an investigation to examine how the properties of the compression springs influence the distance the toy car travels on the race track.

Although the task was developed prior to the release of the NGSS, it was designed to be aligned with *A Framework for K-12 Science Education: Practices, Crosscutting Concepts, and Core Ideas*. The task is related to the crosscutting concept of "energy and matter: flows, cycles and conservation." The task was designed to be aligned with two scientific practices: planning an investigation and analyzing and interpreting data. The concepts are introduced to students by pro-

---

[6]Text is adapted from Haertel et al. (2012). Used with permission.

viding them with opportunities to track changes in energy and matter into, out of, and within systems. The task targets three disciplinary core ideas: definitions of energy, conservation of energy and energy transfer, and the relationship between energy and force.

Design of the Task

The task was designed using a "design pattern," a tool developed to support work at the step of domain modeling in evidence-centered design, which involves the articulation and coordination of claims and evidence statements (see Mislevy et al., 2003). Design patterns help an assessment developer consider the key elements of an assessment argument in narrative form. The subsequent steps in the approach build on the arguments sketched out in domain modeling and represented in the design patterns, including designing tasks to obtain the relevant evidence, scoring performance, and reporting the outcomes. The specific design pattern selected for this task supports the writing of storyboards and items that address scientific reasoning and process skills in planning and conducting experimental investigations. This design pattern could be used to generate task models for groups of tasks for science content strands that are amenable to experimentation.

In the design pattern, the relevant knowledge, skills, and abilities (i.e., the claims about student competence) assessed for this task include the following (Rutstein and Haertel, 2012):

- ability to identify, generate, or evaluate a prediction/hypothesis that is testable with a simple experiment;
- ability to plan and conduct a simple experiment step-by-step given a prediction or hypothesis;
- ability to recognize that at a basic level, an experiment involves manipulating one variable and measuring the effect on (or value of) another variable;
- ability to identify variables of the scientific situation (other than the ones being manipulated or treated as an outcome that should be controlled (i.e., kept the same) in order to prevent misleading information about the nature of the causal relationship; and
- ability to interpret or appropriately generalize the results of a simple experiment or to formulate conclusions or create models from the results.

Evidence of these knowledge, skills, and abilities will include both observations and work products. The potential observations include the following (Rutstein and Haertel, 2012):

- Generate a prediction/hypothesis that is testable with a simple experiment.
- Provide a "plausibility" (explanation) of plan for repeating an experiment.
- Correctly identify independent and dependent variables.
- Accurately identify variables (other than the treatment variables of interest) that should be controlled or made equivalent (e.g., through random assignment).
- Provide a "plausibility" (explanation) of design for a simple experiment.
- Be able to accurately critique the experimental design, methods, results, and conclusions of others.
- Recognize data patterns from experimental data.

The relevant work products include the following (Rutstein and Haertel, 2012):

- Select, identify, or evaluate an investigable question.
- Complete some phases of experimentation with given information, such as selection levels or determining steps.
- Identify or differentiate variables that do and do not need to be controlled in a given scientific situation.
- Generate an interpretation/explanation/conclusion from a set of experimental results.

The Pinball Car Task[7]

Scene 1: A student poses a hypothesis that can be investigated using the simulation presented in the task. The student is introduced to the task and provided with some background information that is important throughout the task: see Figure 3-3. Science terminology and other words that may be new to the student (highlighted in bold) have a roll-over feature that shows their definition when the student scrolls over the word.

The student selects three of nine compression springs to be used in the pinball plunger and initiates a simulation, which generates a table of data that illustrates how far the race car traveled on the race track using the particular compression springs that were selected. Data representing three trial runs are presented

---

[7]This section is largely taken from Rutstein and Haertel (2012).

To demonstrate the difference between **potential energy** and **kinetic energy** your science teacher has set up a Pinball Car race for the class. The race uses toy cars that are powered by a **spring-loaded plunger**.

Each student must use the same car on the same track. The goal is to pick a spring which will make the car go the furthest distance.

| | Once you click next you cannot go back | |

FIGURE 3-3 Task introduction.

NOTE: See text for discussion.

SOURCE: Rutstein and Haertel (2012, Appendix A2). Reprinted with permission from SRI International.

Click the button to play the animation of the Pinball Car race. Use the frame number to answer the questions on the right.

Time: 1   2   3   4   5   6

1. As the plunger is pulled back and released, in what time segment does the spring have the greatest amount of potential energy?
[Select menu for 1 – 6]

2. As the plunger is pulled back and released, in what time segment does the spring have the greatest amount of kinetic energy?
[Select menu for 1 – 6]

| | Once you click next you cannot go back | |

FIGURE 3-4 Animation of a pinball car race.

NOTE: See text for discussion.

SOURCE: Rutstein and Haertel (2012, Appendix A2). Reprinted with permission from SRI International.

each time the simulation is initiated. The student runs the simulation twice for a total of six trials of data for each of the three springs selected.

Scene 2: The student plays an animation that shows what a pinball car race might look like in the classroom: see Figure 3-4. The student uses the animation and its time code to determine the point in which the spring had the greatest potential and kinetic energy.

Scene 3: This scene provides students with background information about springs and introduces them to two variables, the number of coils and the thickness of the wire: see Figure 3-5.

Scene 4: Using the information from Scene 3, the student poses a hypothesis about how these properties might influence the distance the race car travels after the spring plunger is released; see Figure 3-6. The experiment requires that students vary or control each of the properties of the spring.

Background Information	Progress bar	

| You learn that different springs have different amounts of potential energy. After looking through the collection of springs available, you notice that the springs differ on two properties.

These properties might affect the amount of potential energy each spring can store which would then affect the distance the car could travel. | These properties are:

• The number of coils in the spring
  • Some springs have **a lot** of coils
  • Some springs have **some** coils
  • Some springs have **a few** coils

• Thickness of the wire
  • Some springs have **very thick** wire
  • Some springs have **moderately thick** wire
  • Some springs have **thin** wire |

	Once you click next you cannot go back	

FIGURE 3-5 Background information.

NOTE: See text for discussion.

SOURCE: Rutstein and Haertel (2012, Appendix A2). Reprinted with permission from SRI International.

Developing Assessments for the Next Generation Science Standards

Number of Coils	Thickness of the wire	3. Pick a hypothesis to test by filling in the blanks in the statement below:
_____	_____	**I hypothesize that the distance the car travels is affected by the** ___ (number of coils/diameter of the coils).
o A lot	o Very thick	**I believe that the car would travel farther with** ___ (more coils/less coils, coils with larger diameter, coils with smaller diameter).
o Some	o Moderately thick	
o A few	o Thin	

You may click on the options above to see the spring that would be used in the Pinball Car race.	4. Explain your hypothesis: What effect does the ___ (FILL in answer from above such as number of coils) ___ have on the potential and kinetic energy of the spring?

Once you click next you cannot go back

FIGURE 3-6 Picking a hypothesis.

NOTE: See text for discussion.

SOURCE: Rutstein and Haertel (2012, Appendix A2). Reprinted with permission from SRI International.

Your hypothesis is: [FILL IN PREVIOUS ANSWER] springs with ___ (more coils/less coils/larger diameter/smaller diameter) will make the car go further.

You will now design an experiment to test this hypothesis.
Select three springs to use in your experiment.
For each spring, choose the number of coils and a thickness. The springs can have the same settings or you can vary the springs by selecting different settings for one or more variables.

What settings you choose for each of these variables should be based on your hypothesis.

5.	Spring 1 [picture]	Spring 2 [picture]	Spring 3 [picture]
Number of Coils	o A lot o Some o A few	o A lot o Some o A few	o A lot o Some o A few
Thickness of the wire	o Very Thick o Moderately Thick o Thin	o Very Thick o Moderately Thick o Thin	o Very Thick o Moderately Thick o Thin

6. Explain why the choices you made for the settings of the springs are appropriate for your hypothesis.

Once you click next you cannot go back

FIGURE 3-7 Designing an experiment for the hypothesis.

NOTE: See text for discussion.

SOURCE: Rutstein and Haertel (2012, Appendix A2). Reprinted with permission from SRI International.

Scene 5: The student decides whether one or both of the properties of the spring will serve as independent variables and whether one or more of the variables will serve as control variables; see Figure 3-7.

Scene 6: In completing the task, the student decides how many trials of data are needed to produce reliable measurements and whether the properties of the springs need to be varied and additional data collected before the hypothesis can be confirmed or disconfirmed.

Scene 7: Once a student has decided on the levels of the properties of the spring to be tested, the simulation produces a data table, and the student must graph the data and analyze the results.

Scene 8: Based on the results, the student may revise the hypothesis and run the experiment again, changing the settings of the variables to reflect a revision of their model of how the properties of the springs influence the distance the toy car travels: see Figure 3-8.

Questions 9 and 10 out of 12	Progress bar	

Your hypothesis is: [FILL IN PREVIOUS ANSWER] springs with ___ (more coils/less coils/larger diameter/smaller diameter) will make the car go further.

You have an opportunity to run your experiment again to obtain more information about your hypothesis. You can either change the settings of your springs or leave them the same. Remember that you are still testing the same hypothesis and so your settings for the spring must reflect this hypothesis.	9.	Spring 1 [picture]	Spring 2 [picture]	Spring 3 [picture]

	Spring 1	Spring 2	Spring 3
Number of Coils	o A lot o Some o A few	o A lot o Some o A few	o A lot o Some o A few
Thickness of the wire	o Very Thick o Moderately Thick o Thin	o Very Thick o Moderately Thick o Thin	o Very Thick o Moderately Thick o Thin

10. Explain why the choices you made for the settings of the springs are appropriate for your hypothesis.

Once you click next you cannot go back →

FIGURE 3-8 Option to rerun the experiment.

NOTE: See text for discussion.

SOURCE: Rutstein and Haertel (2012, Appendix A2). Reprinted with permission from SRI International.

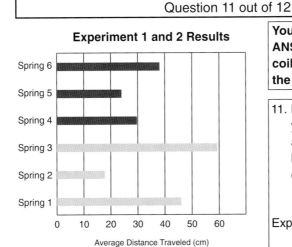

Experiment 1 and 2 Results

Your hypothesis is: [FILL IN PREVIOUS ANSWER] Springs with ___ (more coils/less coils/thicker wire/thinner wirer) will make the car go further.

11. How do the results from Trial 2 relate to your hypothesis:
   a) These results **support** my hypothesis
   b) These results **contradict** my hypothesis
   c) These results **do not provide information** about my hypothesis.

Explain your answer:

Once you click next you cannot go back ⇨

FIGURE 3-9 Results of two experiments.

NOTE: See text for discussion.

SOURCE: Rutstein and Haertel (2012, Appendix A2). Reprinted with permission from SRI International.

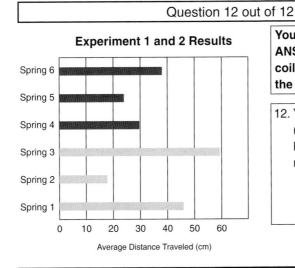

Experiment 1 and 2 Results

Your hypothesis is: [FILL IN PREVIOUS ANSWER] Springs with ___ (more coils/less coils/thicker wire/thinner wirer) will make the car go further.

12. You said that the results from Experiment 1 (fill in from item 7) your hypothesis. How did Experiment 2 help you learn more about your hypothesis?

Once you click next you cannot go back ⇨

FIGURE 3-10 Use of results from the two experiments.

NOTE: See text for discussion.

SOURCE: Rutstein and Haertel (2012, Appendix A2). Reprinted with permission from SRI International.

After running a few more experiments you decide to pick a
spring with a small number of coils and thick wire. You end up
winning the race and your science teacher is very impressed!

FIGURE 3-11 Final result of the pinball car task.

NOTE: See text for discussion.

SOURCE: Rutstein and Haertel (2012, Appendix A2). Reprinted with permission from SRI
International.

Scene 9: If the student chose to run the experiment a second time, the
results of both experiments are now shown on the same bar chart: see Figure 3-9.

Scene 10: The student is asked how the results of the second experiment
relate to her or his hypothesis: see Figure 3-10.

Scene 11: The final scene gives the student the spring characteristics that
would lead to the car going the furthest distance and winning the race: see Figure
3-11.

Scoring

The pinball car task was developed as a prototype to demonstrate the use of
design patterns in developing technology-enhanced, scenario-based tasks of hard-
to-assess concepts. It has been pilot tested but not administered operationally. The
developers suggest that the tasks could be scored several ways. It could be scored
by summing those items aligned primarily to content standards and those aligned
primarily to practice standards, thus producing two scores. Or the task could
generate an overall score based on the aggregation of all items, which is more in
keeping with the idea of three-dimensional science learning in the framework.

Alternatively, the specific strengths and weaknesses in students' understanding could be inferred from the configurations of their correct and incorrect responses according to some more complex decision rule.

Construct Modeling: Measuring Silkworms

In this task, 3rd-grade elementary school students explored the distinction between organismic and population levels of analysis by inventing and revising ways of visualizing the measures of a large sample of silkworm larvae at a particular day of growth. The students were participating in a teacher-researcher partnership aimed at creating a multidimensional learning progression to describe practices and disciplinary ideas that would help young students consider evolutionary models of biological diversity.

The learning progression was centered on student participation in the invention and revision of representations and models of ecosystem functioning, variability, and growth at organismic and population levels (Lehrer and Schauble, 2012). As with other examples in this report, the task was developed prior to the publication of the NGSS, but is aligned with the life sciences progression of the NGSS: see Tables 1-1 and 3-1. The practices listed in the tables were used in the

TABLE 3-1 Assessment Targets for Example 3 (Measuring Silkworms) and the Next Generation Science Standards (NGSS) Learning Progressions

Disciplinary Core Idea from the NGSS	Practices	Performance Expectation	Crosscutting Concept
LS1.A Structure and function (grades 3-5): Organisms have macroscopic structures that allow for growth.	Asking questions Planning and carrying out Investigations Analyzing and interpreting data	Observe and analyze the external structures of animals to explain how these structures help the animals meet their needs.	Patterns
LS1.B Growth and development of organisms (grades 3-5): Organisms have unique and diverse life cycles.	Using mathematics Constructing explanations Engaging in argument from evidence Communicating information	Gather and use data to explain that young animals and plants grow and change. Not all individuals of the same kind of organism are exactly the same: there is variation.	

NOTES: LS1.A and LS1.B refer to the disciplinary core ideas in the framework: see Box 2-1 in Chapter 2.

development of core ideas about organism growth. The classroom-embedded task was designed to promote a shift in student thinking from the familiar emphasis on individual organisms to consideration of a population of organisms: to do so, the task promotes the practice of analyzing and interpreting data. Seven dimensions have been developed to specify this multidimensional construct, but the example focuses on just one: reasoning about data representation (Lehrer et al., 2013). Hence, an emerging practice of visualizing data was coordinated with an emerging disciplinary core idea, population growth, and with the crosscutting theme of pattern.

The BEAR Assessment System for Assessment Design

The BEAR Assessment System (BAS) (Wilson, 2005) is a set of practical procedures designed to help one apply the construct-modeling approach. It is based on four principles—(1) a developmental perspective, (2) a match between instruction and assessment, (3) management by teachers, and (4) evidence of high quality—each of which has a corresponding element: see Figure 3-12. These elements func-

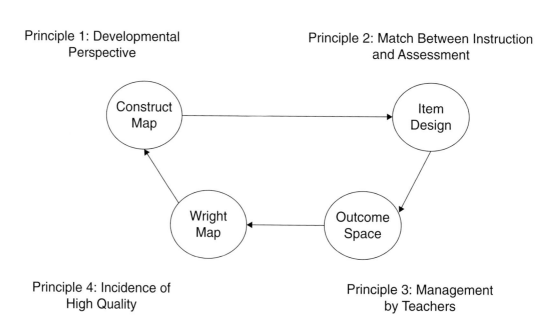

FIGURE 3-12 The BEAR system.
SOURCE: Wilson (2009, fig. 2, p. 718). Reprinted with permission from John Wiley & Sons.

Developing Assessments for the Next Generation Science Standards

tion in a cycle, so that information gained from each phase of the process can be used to improve other elements. Current assessment systems rarely allow for this sort of continuous feedback and refinement, but the developers of the BAS believe it is critical (as in any engineering system) to respond to results and developments that could not be anticipated.

The first element of BAS is the *construct map*, which defines what is to be assessed. The construct map has been described as a visual metaphor for the ways students' understanding develops, and, correspondingly, how it is hypothesized that their responses to items might change (Wilson, 2005). Figure 3-13 is an example of a construct map for one aspect of analyzing and interpreting data, data display (abbreviated as "DaD"). The construct map describes significant milestones in children's reasoning about data representation, presenting them as a progression from a stage in which students focus on individual case values (e.g., the students describe specific data points) to a stage when they are capable of reasoning about patterns of aggregation. The first and third columns of Figure 3-13 display the six levels associated with this construct, with Level 6 being the most sophisticated.

The second BAS element is *item design,* which specifies how the learning performances described by the construct will be elicited. It is the means by which the match between the curriculum and the assessment is established. Item design can be described as a set of principles that allow one to observe students under a set of standard conditions (Wilson, 2005). Most critical is that the design specifications make it possible to observe each of the levels and sublevels described in the construct map.

The third element, *outcome space,* is a general guide to the way students' responses to items developed in relation to a particular construct map will be valued. The more specific guidance developed for a particular item is used as the actual scoring guide, which is designed to ensure that student responses can be interpreted in light of the construct map. The third column of Figure 3-13 is a general scoring guide. The final element of BAS, a *Wright map*, is a way to apply the measurement model, to collect the data and link it back to the goals for the assessment and the construct maps.[8] The system relies on a multidimensional way of organizing statistical evidence of the quality of the assessment, such as its reli-

---

[8] A Wright map is a figure that shows both student locations and item locations on the same scale—distances along it are interpreted in terms of the probability of success of a student at that location succeeding at an item at that location (see Wilson, 2004, Chapter 5)

Level		Performances		Examples
DaD6 Integrate case with aggregate perspectives.	DaD6 A	Discuss how general patterns or trends are either exemplified or missing from subsets of cases.		• Relate qualities of a case as an example of general qualities of a region of data (case as typical of data region). • Notice that a subset of cases does not seem to fit the trends observed or conjectured.
DaD5 Consider the data in aggregate when interpreting or creating displays.	DaD5 B	Quantify aggregate property of the display using one or more of the following: ratio, proportion or percent.		• "I found out that measurements between 45 and 55 were 70% of our measurements. So, I guess the true height is somewhere between 45 and 55." • Students annotate their display to show percentages within particular regions.
	DaD5 A	Recognize that a display provides information about the data as a collective.		• "The distribution of the data is wider for rounded-nosecone rockets than for pointed-nosecone rockets. Maybe that's because pointed rockets flights are more consistent." • "When we measure different things, we keep getting a bell shape. That's because we tend to get around the real measure most of the time, but sometimes we make big mistakes."
DaD4 Recognize or apply scale properties to the data.	DaD4 B	Recognize the effects of changing bin size on the shape of the distribution.		• "If we make bin size wider, the tower in the center will pop up."
	DaD4 A	Display data in ways that use its continuous scale (when appropriate) to see holes and clumps in the data.		- "Number line" display: - "Bin" display:

Level		Performances	Examples
DaD3 Notice or construct groups of similar values.	DaD3 A	Notice or construct groups of similar values from distinct values.	• Create unordered bins, and comment on, for example, the number of occurrences of 40s vs. the number of 50s. • When asked to name bins in a preset display, assigns discontinuous and/or unequal intervals to the bins, such as 2-25, 26-36, 37-45. • Create equal interval bins but leave out intermediate intervals. • Notice "plateaus" in the case display or a group of similar values. • "This number, 193, is really different, because the others are all between 160 and 165." • "Most of batteries lasted between 120 to 140 minutes."
DaD2 Interpret and/or produce data displays as all collections of individual cases.	DaD2 B	Construct/interpret data by considering ordinal properties.	• "The data start out with the lowest measurement and go to the highest one." • Create display by ordering data as a list or case-value graph.
	DaD2 A	Concentrate on specific data points without relating these to any structure in the data.	• Identify maximum and minimum values. • "The only thing I can tell is this (193) is the highest." • "154 is the number in the middle of the list (without ordering the data)." • "This number is the biggest."
DaD1 Create displays or interpret displays without reference to goals of data creation.	DaD1 A	Create or interpret data displays without relating to the goals of the inquiry.	• "We grouped even and odd numbers because we like even and odd numbers." • "I put these two values (19 and 11) on the top because that's my birthday - Nov. 19th!" • "This display has lots of numbers."

FIGURE 3-13 A construct map of the data display (DaD) construct.

NOTE: See text for discussion.

SOURCE: Wilson et al. (2013). Copyright by the author; used with permission.

ability, validity, and fairness. Item-response models show students' performance on particular elements of the construct map across time; they also allow for comparison within a cohort of students or across cohorts.

The Silkworm Growth Activity

In our example, the classroom activity for assessment was part of a classroom investigation of the nature of growth of silkworm larvae. The silkworm larvae are a model system of metamorphic insect growth. The investigation was motivated by students' questions and by their decisions about how to measure larvae at different days of growth. The teacher asked students to invent a display that communicated what they noticed about the collection of their measures of larvae length on a particular day of growth.

Inventing a display positioned students to engage spontaneously with the forms of reasoning described by the DaD construct map (see Figure 3-13, above): the potential solutions were expected to range from Levels 1 to 5 of the construct. In this classroom-based example, the item design is quite informal, being simply what the teacher asked the students to do. However, the activity was designed to support the development of the forms of reasoning described by the construct.

One data display that several groups of students created was a case-value graph that ordered each of 261 measurements of silkworms by magnitude: see Figure 3-14. The resulting display occupied 5 feet of the classroom wall. In this representation, the range of data is visible at a glance, but the icons resembling the larvae and representing each millimeter of length are not uniform. This is an example of student proficiency at Level 2 of the construct map. The second display developed by the student groups used equal-sized intervals to show equivalence among classes of lengths: see Figure 3-15. By counting the number of cases within each interval, the students made a center clump visible. This display makes the *shape* of the data more visible; however, the use of space was not uniform and produced some misleading impressions about the frequency of longer or shorter larvae. This display represents student proficiency at Level 3 of the construct map.

The third display shows how some students used the measurement scale and counts of cases, but because of difficulties they experienced with arranging the display on paper, they curtailed all counts greater than 6: see Figure 3-16. This display represents student proficiency at Level 4 of the construct map

The displays that students developed reveal significant differences in how they thought about and represented their data. Some focused on case values, while others were able to use equivalence and scale to reveal characteristics of the data

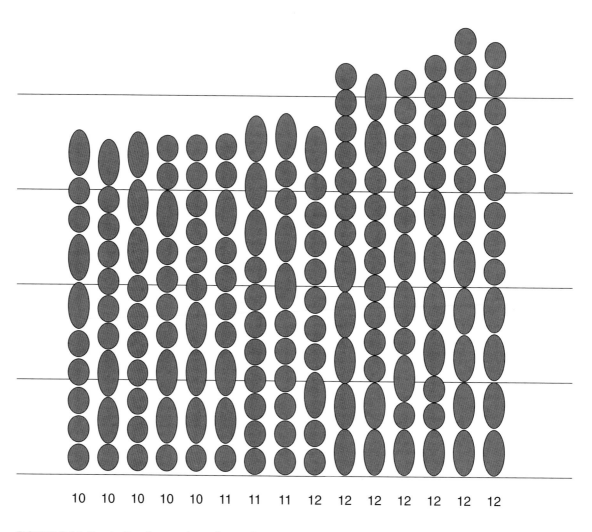

FIGURE 3-14 Facsimile of a portion of a student-created case-value representation of silkworm larvae growth.

SOURCE: Lehrer (2011). Copyright by the author; used with permission.

in aggregate. The construct map helped the teacher appreciate the significance of these differences.

To help students develop their competence at representing data, the teacher invited them to consider what selected displays show and do not show about the data. The purpose was to convey that all representational choices emphasize certain features of data and obscure others. During this conversation, the students critiqued how space was used in the displays to represent lengths of the larvae and began to

Developing Assessments for the Next Generation Science Standards

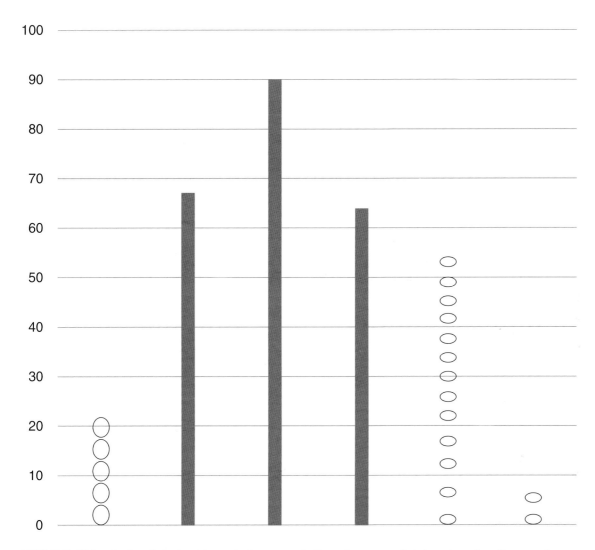

FIGURE 3-15 Facsimile of student-invented representation of groups of data values for silkworm larvae growth.

NOTE: The original used icons to represent the organisms in each interval.

SOURCE: Lehrer (2011). Copyright by the author; used with permission.

appreciate the basis of conventions about display regarding the use of space, a form of meta-representational competence (diSessa, 2004). The teacher also led a conversation about the mathematics of display, including the use of order, count, and interval and measurement scale to create different senses of the shape of the data. (This focus on shape corresponds to the crosscutting theme of pattern in the

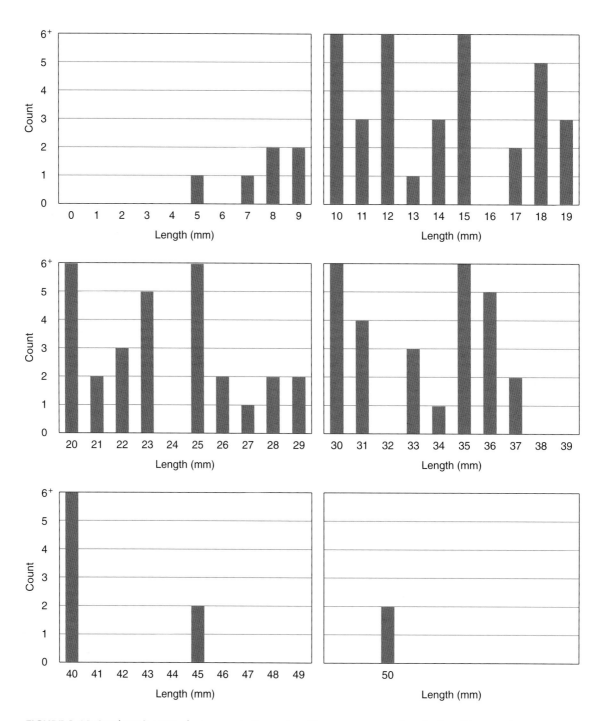

FIGURE 3-16 Student-invented representation using the measurement scale for silkworm larvae growth.

SOURCE: Lehrer (2011). Copyright by the author; used with permission.

Developing Assessments for the Next Generation Science Standards

NGSS.) Without this instructional practice, well-orchestrated discussion led by the teacher—who was guided by the construct map in interpreting and responding to student contributions—students would be unlikely to discern the bell-like shape that is often characteristic of natural variation.

The focus on the shape of the data was a gentle introduction to variability that influenced subsequent student thinking about larval growth. As some students examined Figure 3-16, they noticed that the tails of the distribution were comparatively sparse, especially for the longer silkworm larvae, and they wondered why. They speculated that this shape suggested that the organisms had differential access to resources. They related this possibility to differences in the timing of larval hatching and conjectured that larvae that hatched earlier might have begun eating and growing sooner and therefore acquired an advantage in the competition for food. The introduction of competition into their account of variability and growth was a new form of explanation, one that helped them begin to think beyond individual organisms to the population level. In these classroom discussions, the teacher blends instruction and diagnosis of student thinking for purposes of formative assessment.

Other Constructs and a Learning Progression

Our example is a classroom-intensive context, and formal statistical modeling of this small sample of particular student responses would not be useful. However, the responses of other students involved in learning about data and statistics by inventing displays, measures, and models of variability (Lehrer et al., 2007, 2011) were plotted using a DaD construct map (see Figure 3-13, above), and the results of the analysis of those data are illustrated in Figure 3-17 (Schwartz et al., 2011). In this figure, the left-hand side shows the units of the scale (in logits[9]) and also the distributions of the students along the DaD construct. The right-hand side shows the locations of the items associated with the levels of the construct—the first column (labeled "NL") is a set of responses that are pre-Level 1—that is, they are responses that do not yet reach Level 1, but they show some relevancy, even if it is just making appropriate reference to the item. These points (locations of the thresholds) are where a student is estimated to have a probability of 0.50 of

---

[9]The logit scale is used to locate both examinees and assessment tasks relative to a common, underlying (latent) scale of both student proficiency and task difficulty. The difference in logits between an examinee's proficiency and a task's difficulty is equal to the logarithm of the odds of a correct response to that task by that examinee, as determined by a statistical model.

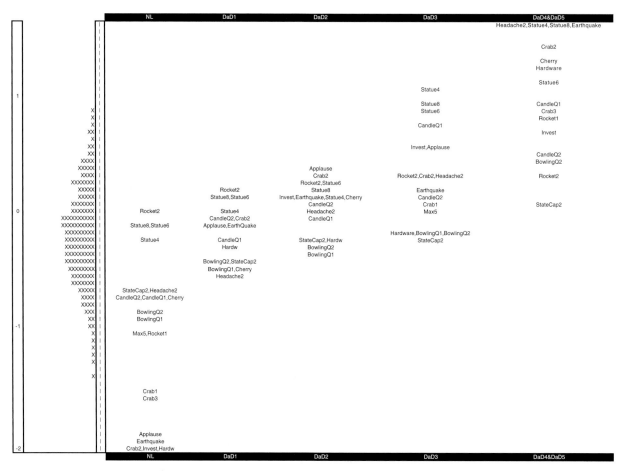

FIGURE 3-17 Wright map of the DaD construct.

SOURCE: Wilson et al. (2013). Copyright by the author; used with permission.

responding at that level or below. Using this figure, one can then construct bands that correspond to levels of the construct and help visualize relations between item difficulties and the ordered levels of the construct. This is a more focused test of construct validity than traditional measures of item fit, such as the mean square or others (Wilson, 2005).

The DaD construct is but one of seven assessed with this sample of students, so BAS was applied to each of the seven constructs: theory of measurement, DaD, meta-representational competence, conceptions of statistics, chance, models of variability, and informal inference (Lehrer et al., 2013); see Figure 3-18.

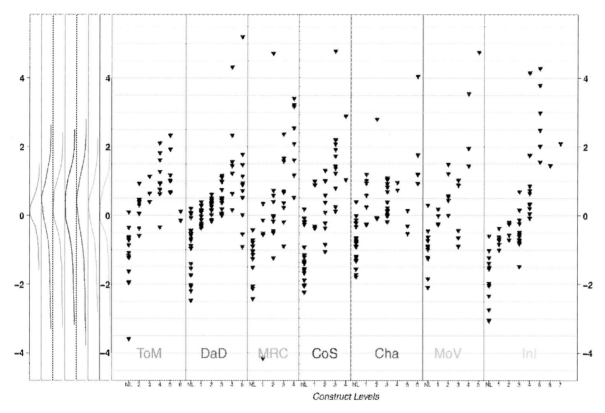

FIGURE 3-18 Wright map of the seven dimensions assessed for analyzing and interpreting data.

NOTES: Cha = chance, CoS = conceptions of statistics, DaD = data display, InI = informal inference, MoV = models of variability, MRC = meta-representational competence, ToM = theory of measurement. See text for discussion.

SOURCE: Wilson et al. (2013). Copyright by the author; used by permission.

1. *Theory of measurement* maps the degree to which students understand the mathematics of measurement and develop skills in measuring. This construct represents the basic area of knowledge in which the rest of the constructs are played out.

2. *DaD* traces a progression in learning to construct and read graphical representations of the data from an initial emphasis on cases toward reasoning based on properties of the aggregate.

3. *Meta-representational competence*, which is closely related to DaD, proposes keystone performances as students learn to harness varied representations for making claims about data and to consider tradeoffs among representations in light of these claims.

4. *Conceptions of statistics* propose a series of landmarks as students come to first recognize that statistics measure qualities of the distribution, such as center and spread, and then go on to develop understandings of statistics as generalizable and as subject to sample-to-sample variation.

5. *Chance* describes the progression of students' understanding about how chance and elementary probability operate to produce distributions of outcomes.

6. *Models of variability* refer to the progression of reasoning about employing chance to model a distribution of outcomes produced by a process.

7. *Informal inference* describes a progression in the basis of students' inferences, beginning with reliance on cases and ultimately culminating in using models of variability to make inferences based on single or multiple samples.

These seven constructs can be plotted as a learning progression that links the theory of measurement, a construct that embodies a core idea, with the other six constructs, which embody practices: see Figure 3-19. In this figure, each vertical set of levels is one of the constructs listed above. In addition to the obvious links between the levels within a construct, this figure shows hypothesized links between specific levels of different constructs. These are interpreted as necessary prerequisites: that is, the hypothesis is that a student needs to know the level at the base of the arrow before he or she can succeed on the level indicated at the point of the arrow. The area labeled as "bootstrapping" is a set of levels that require mutual support. Of course, performance on specific items will involve measurement error, so these links need to be investigated using multiple items within tasks.

VALIDATION

Despite all the care that is taken in assessment design to ensure that the developed tasks measure the intended content and skills, it is still necessary to evaluate empirically that the inferences drawn from the assessment results are valid. Validity refers to the extent to which assessment tasks measure the skills that they are intended to measure (see, e.g., Kane, 2006, 2013; Messick, 1993; National Research Council, 2001, 2006). More formally, "Validity is an integrated evaluative judgment of the degree to which empirical evidence and theoretical rationales support the adequacy and appropriateness of inferences and actions based on the test" (Messick, 1989, p. 13). Validation involves evaluation of the proposed interpretations and uses of the assessment results, using different kinds of evidence,

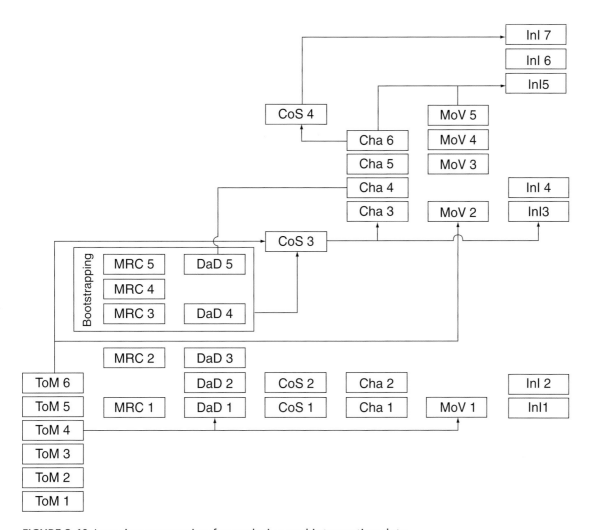

FIGURE 3-19 Learning progression for analyzing and interpreting data.

NOTES: Cha = chance, CoS = conceptions of statistics, DaD = data display, InI = informal inference, MoV = models of variability, MRC = meta-representational competence, ToM = theory of measurement. See text for discussion.

evidence that is rational and empirical and both qualitative and quantitative. For the examples discussed in this report, validation would include analysis of the processes and theory used to design and develop the assessment, evidence that the respondents were indeed thinking in the ways envisaged in that theory, the internal structure of the assessment, the relationships between results and other outcome measures, and whether the consequences of using the assessment results were as expected, and other studies designed to examine the extent to which the intended interpretations of assessment results are fair, justifiable, and appropriate for a given purpose (see American Educational Research Association, American Psychological Association, and National Council on Measurement in Education, 1999).

Evidence of validity is typically collected once a preliminary set of tasks and corresponding scoring rubrics have been developed. Traditionally, validity concerns associated with achievement tests have focused on test content, that is, the degree to which the test samples the subject matter domain about which inferences are to be drawn. This sort of validity is confirmed through evaluation of the alignment between the content of the assessment tasks and the subject-matter framework, in this case, the NGSS.

Measurement experts increasingly agree that traditional external forms of validation, which emphasize consistency with other measures, as well as the search for indirect indicators that can show this consistency statistically, should be supplemented with evidence of the cognitive and substantive aspects of validity (Linn et al., 1991; Messick, 1993). That is, the trustworthiness of the interpretation of test scores should rest in part on empirical evidence that the assessment tasks actually reflect the intended cognitive processes. There are few alternative measures that assess the three-dimensional science learning described in the NGSS and hence could be used to evaluate consistency, so the empirical validity evidence will be especially important for the new assessments that states will be developing as part of their implementation of the NGSS.

Examining the processes that students use as they perform an assessment task is one way to evaluate whether the tasks are functioning as intended, another important component of validity. One method for doing this is called *protocol analysis* (or *cognitive labs*), in which students are asked to think aloud as they solve problems or to describe retrospectively how they solved the problem (Ericsson and Simon, 1984). Another method is called *analysis of reasons*, in which students are asked to provide rationales for their responses to the tasks. A third method, *analysis of errors*, is a process of drawing inferences about students' processes from incorrect procedures, concepts, or representations of the problems (National Research Council, 2001).

The empirical evidence used to investigate the extent to which the various components of an assessment actually perform together in the way they were designed to is referred to collectively as evidence based on the *internal structure* of the test (see American Educational Research Association, American Psychological Association, and National Council on Measurement in Education, 1999). For example, in our example of measuring silkworm larvae growth, one form of evidence based on internal structure would be the match between the hypothesized levels of the construct maps and the empirical difficulty order shown in the measurement map in Figure 3-15 above.

One critical aspect of validity is fairness. An assessment is considered fair if test takers can demonstrate their proficiency in the targeted content and skills without other, irrelevant factors interfering with their performance. Many attributes of test items can contribute to what measurement experts refer to as *construct-irrelevant variance*, which occurs when the test questions require skills that are not the focus of the assessment. For instance, an assessment that is intended to measure a certain science practice may include a lengthy reading passage. Besides assessing skill in the particular practice, the question will also require a certain level of reading skill. Assessment respondents who do not have sufficient reading skills will not be able to accurately demonstrate their proficiency with the targeted science skills. Similarly, respondents who do not have a sufficient command of the language in which an assessment is presented will not be able to demonstrate their proficiency in the science skills that are the focus of the assessment. Attempting to increase fairness can be difficult, however, and can create additional problems. For example, assessment tasks that minimize reliance on language by using online graphic representations may also introduce a new construct-irrelevant issue because students have varying familiarity with these kinds of representations or with the possible ways to interact with them offered by the technology.

Cultural, racial, and gender issues may also pose fairness questions. Test items should be designed so that they do not in some way disadvantage the respondent on the basis of those characteristics, social economic status, or other background characteristics. For example, if a passage uses an example more familiar or accessible to boys than girls (e.g., an example drawn from a sport in which boys are more likely to participate), it may give the boys an unfair advantage. Conversely, the opposite may occur if an example is drawn from cooking (with which girls are more likely to have experience). The same may happen if the material in the task is more familiar to students from a white, Anglo-Saxon background than to students from minority racial and ethnic backgrounds or more familiar to students who live in urban areas than those in rural areas.

It is important to keep in mind that attributes of tasks that may seem unimportant can cause differential performance, often in ways that are unexpected and not predicted by assessment designers. There are processes for bias and sensitivity reviews of assessment tasks that can help identify such problems before the assessment is given (see, e.g., Basterra et al., 2011; Camilli, 2006; Schmeiser and Welch, 2006; Solano-Flores and Li, 2009). Indeed this process was begun by the NGSS. Their development work included a process to review and refine the performance expectations using this lens (see Appendix 4 of the NGSS). After an assessment has been given, analyses of differential item functioning can help identify problematic questions so that they can be excluded from scoring (see, e.g., see Camilli and Shepard, 1994; Holland and Wainer, 1993; Sudweeks and Tolman, 1993).

A particular concern for science assessment is the opportunity to learn—the extent to which students have had adequate instruction in the assessed material to be able to demonstrate proficiency on the targeted content and skills. Inferences based on assessment results cannot be valid if students have not had the opportunity to learn the tested material, and the problem is exacerbated when access to adequate instruction is uneven among schools, districts, and states. This equity issue has particular urgency in the context of a new approach to science education that places many new kinds of expectations on students. The issue was highlighted in *A Framework for K-12 Science Education: Practices, Crosscutting Concepts, and Core Ideas* (National Research Council, 2012a, p. 280), which noted:

. . . *access to high quality education in science and engineering is not equitable across the country; it remains determined in large part by an individual's socioeconomic class, racial or ethnic group, gender, language background, disability designation, or national origin.*

The validity of science assessments designed to evaluate the content and skills depicted in the framework could be undermined simply because students do not have equal access to quality instruction. As noted by Pellegrino (2013), a major challenge in the validation of assessments designed to measure the NGSS performance expectations is the need for such work to be done in instructional settings where students have had adequate opportunity to learn the integrated knowledge envisioned by the framework and the NGSS. We consider this issue in more detail in Chapter 7 in the context of suggestions regarding implementation of next generation science assessments.

CONCLUSION AND RECOMMENDATION

CONCLUSION 3-1 Measuring three-dimensional learning as conceptualized in the framework and the Next Generation Science Standards (NGSS) poses a number of conceptual and practical challenges and thus demands a rigorous approach to the process of designing and validating assessments. The endeavor needs to be guided by theory and research about science learning to ensure that the resulting assessment tasks are (1) consistent with the framework and NGSS, (2) provide information to support the intended inferences, and (3) are valid for the intended use.

RECOMMENDATION 3-1 To ensure that assessments of a given performance expectation in the Next Generation Science Standards provide the evidence necessary to support the intended inference, assessment designers should follow a systematic and principled approach to assessment design, such as evidence-centered design or construct modeling. In so doing, multiple forms of evidence need to be assembled to support the validity argument for an assessment's intended interpretive use and to ensure equity and fairness.

4

CLASSROOM ASSESSMENT

$\mathbf{A}$ssessments can be classified in terms of the way they relate to instructional activities. The term classroom assessment (sometimes called internal assessment) is used to refer to assessments designed or selected by teachers and given as an integral part of classroom instruction. They are given during or closely following an instructional activity or unit. This category of assessments may include teacher-student interactions in the classroom, observations, student products that result directly from ongoing instructional activities (called "immediate assessments"), and quizzes closely tied to instructional activities (called "close assessments"). They may also include formal classroom exams that cover the material from one or more instructional units (called "proximal assessments").[1] This category may also include assessments created by curriculum developers and embedded in instructional materials for teacher use.

In contrast, external assessments are designed or selected by districts, states, countries, or international bodies and are typically used to audit or monitor learning. External assessments are usually more distant in time and context from instruction. They may be based on the content and skills defined in state or national standards, but they do not necessarily reflect the specific content that was covered in any particular classroom. They are typically given at a time that is determined by administrators, rather than by the classroom teacher. This category includes such assessments as the statewide science tests required by the No Child

---

[1]This terminology is drawn from Ruiz-Primo et al. (2002) and Pellegrino (2013).

Left Behind Act or other accountability purposes (called "distal assessments"), as well as national and international assessments: the National Assessment of Educational Progress and the Programme for International Student Assessment (called "remote assessments"). Such external assessments and their monitoring function are the subject of the next chapter.

In this chapter, we illustrate the types of assessment tasks that can be used in the classroom to meet the goals of *A Framework for K-12 Science Education: Practices, Crosscutting Concepts, and Core Ideas* (National Research Council, 2012a, hereafter referred to as "the framework") and the *Next Generation Science Standards: For States, By States* (NGSS Lead States, 2013). We present example tasks that we judged to be both rigorous and deep probes of student capabilities and also to be consistent with the framework and the Next Generation Science Standards (NGSS). We discuss external assessments in Chapter 5 and the integration of classroom and external assessments into a coherent system in Chapter 6. The latter chapter argues that an effective assessment system should include a variety of types of internal and external assessments, with each designed to fulfill complementary functions in assessing achievement of the NGSS performance objectives.

Our starting point for looking in depth at classroom assessment is the analysis in Chapter 2 of what the new science framework and the NGSS imply for assessment. We combine these ideas with our analysis in Chapter 3 of current approaches to assessment design as we consider key aspects of classroom assessment that can be used as a component in assessment of the NGSS performance objectives.

ASSESSMENT PURPOSES: FORMATIVE OR SUMMATIVE

Classroom assessments can be designed primarily to guide instruction (formative purposes) or to support decisions made beyond the classroom (summative purposes). Assessments used for formative purposes occur during the course of a unit of instruction and may involve both formal tests and informal activities conducted as part of a lesson. They may be used to identify students' strengths and weaknesses, assist educators in planning subsequent instruction, assist students in guiding their own learning by evaluating and revising their own work, and foster students' sense of autonomy and responsibility for their own learning (Andrade and Cizek, 2010, p. 4). Assessments used for summative purposes may be administered at the end of a unit of instruction. They are designed to provide evidence of achievement that can be used in decision making, such as assigning grades; making promotion or retention decisions; and classifying test takers according to defined performance categories, such as "basic," "proficient," and "advanced" (levels often used in score reporting) (Andrade and Cizek, 2010, p. 3).

The key difference between assessments used for formative purposes and those used for summative purposes is in how the information they provide is to be used: to guide and advance learning (usually while instruction is under way) or to obtain evidence of what students have learned for use beyond the classroom (usually at the conclusion of some defined period of instruction). Whether intended for formative or summative purposes, evidence gathered in the classroom should be closely linked to the curriculum being taught. This does not mean that the assessment must use the formats or exactly the same material that was presented in instruction, but rather that the assessment task should directly address the concepts and practices to which the students have been exposed.

The results of classroom assessments are evaluated by the teacher or sometimes by groups of teachers in the school. Formative assessments may also be used for reflection among small groups of students or by the whole class together. Classroom assessments can play an integral role in students' learning experiences while also providing evidence of progress in that learning. Classroom instruction is the focus of the framework and the NGSS, and it is classroom assessment—which by definition is integral to instruction—that will be the most straightforward to align with NGSS goals (once classroom instruction is itself aligned with the NGSS).

Currently, many schools and districts administer benchmark or interim assessments, which seem to straddle the line between formative and summative purposes (see Box 4-1). They are formative in the sense that they are used for a diagnostic function intended to guide instruction (i.e., to predict how well students are likely to do on the end-of-year tests). However, because of this purpose, the format they use resembles the end-of-year tests rather than other types of internal assessments commonly used to guide instruction (such as quizzes, classroom dialogues, observations, or other types of immediate assessment strategies that are closely connected to instruction). Although benchmark and interim assessments serve a purpose, we note that they are not the types of formative assessments that we discuss in relation to the examples presented in this chapter or that are advocated by others (see, e.g., Black and Wiliam, 2009; Heritage, 2010; Perie et al., 2007). Box 4-1 provides additional information about these types of assessments.

BOX 4-1

BENCHMARK AND INTERIM ASSESSMENTS

Currently, many schools and districts administer benchmark or interim assessments, which they treat as formative assessments. These assessments use tasks that are taken from large-scale tests given in a district or state or are very similar to tasks that have been used in those tests. They are designed to provide an estimate of students' level of learning, and schools use them to serve a diagnostic function, such as to predict how well students will do on the end-of-year tests.

Like the large-scale tests they closely resemble, benchmark tests rely heavily on multiple-choice items, each of which tests a single learning objective. The items are developed to provide only general information about whether students understand a particular idea, though sometimes the incorrect choices in a multiple-choice item are designed to probe for particular common misconceptions. Many such tasks would be needed to provide solid evidence that students have met the performance expectations for their grade level or grade band.

Teachers use these tests to assess student knowledge of a particular concept or a particular aspect of practice (e.g., control of variables), typically after teaching a unit that focuses on specific discrete learning objectives. The premise behind using items that mimic typical large-scale tests is that they help teachers measure students' progress toward objectives for which they and their students will be held accountable and provide a basis for deciding which students need extra help and what the teacher needs to teach again.

CHARACTERISTICS OF NGSS-ALIGNED ASSESSMENTS

Chapter 2 discusses the implications of the NGSS for assessment, which led to our first two conclusions:

- Measuring the three-dimensional science learning called for in the framework and the Next Generation Science Standards requires assessment tasks that examine students' performance of scientific and engineering practices in the context of crosscutting concepts and disciplinary core ideas. To adequately cover the three dimensions, assessment tasks will generally need to contain multiple components (e.g., a set of interrelated questions). It may be useful to focus on individual practices, core ideas, or crosscutting concerts in the various components of an assessment task, but, together, the components need to support inferences about students' three-dimensional science learning as described in a given performance expectation (Conclusion 2-1).

- The Next Generation Science Standards require that assessment tasks be designed so that they can accurately locate students along a sequence of progressively more complex understandings of a core idea and successively more sophisticated applications of practices and crosscutting concepts (Conclusion 2-2).

Students will likely need repeated exposure to investigations and tasks aligned to the framework and the NGSS performance expectations, guidance about what is expected of them, and opportunities for reflection on their performance to develop these proficiencies, as discussed in Chapter 2. The kind of instruction that will be effective in teaching science in the way the framework and the NGSS envision will require students to engage in science and engineering practices in the context of disciplinary core ideas—and to make connections across topics through the crosscutting ideas. Such instruction will include activities that provide many opportunities for teachers to observe and record evidence of student thinking, such as when students develop and refine models; generate, discuss, and analyze data; engage in both spoken and written explanations and argumentation; and reflect on their own understanding of the core idea and the subtopic at hand (possibly in a personal science journal).

The products of such instruction form a natural link to the characteristics of classroom assessment that aligns with the NGSS. We highlight four such characteristics:

1. the use of a variety of assessment activities that mirror the variety in NGSS-aligned instruction;
2. tasks that have multiple components so they can yield evidence of three-dimensional learning (and multiple performance expectations);
3. explicit attention to the connections among scientific concepts; and
4. the gathering of information about how far students have progressed along a defined sequence of learning.

Variation in Assessment Activities

Because NGSS-aligned instruction will naturally involve a range of activities, classroom assessment that is integral to instruction will need to involve a corresponding variation in the types of evidence it provides about student learning. Indeed, the distinction between instructional activities and assessment activities may be blurred, particularly when the assessment purpose is formative. A classroom assessment may be based on a classroom discussion or a group activity in which students explore and respond to each other's ideas and learn as they go through this process.

Science and engineering practices lend themselves well to assessment activities that can provide this type of evidence. For instance, when students are developing and using models, they may be given the opportunity to explain their models and to discuss them with classmates, thus providing the teacher with an opportunity for formative assessment reflection (illustrated in Example 4, below). Student discourse can give the teacher a window into students' thinking and help to guide lesson planning. A classroom assessment may also involve a formal test or diagnostic quiz. Or it may be based on artifacts that are the products of classroom activities, rather than on tasks designed solely for assessment purposes. These artifacts may include student work produced in the classroom, homework assignments (such as lab reports), a portfolio of student work collected over the course of a unit or a school year (which may include both artifacts of instruction as well as results from formal unit and end-of-course tests), or activities conducted using computer technology. A classroom assessment may occur in the context of group work or discussions, as long as the teacher ensures that all the students that need to be observed are in fact active participants. Summative assessments may also take a variety of forms, but they are usually intended to assess each student's independent accomplishments.

Tasks with Multiple Components

The NGSS performance expectations each blend a practice and, in some cases, also a crosscutting idea with an aspect of a particular core idea. In the past, assessment tasks have typically focused on measuring students' understanding of aspects of core ideas or of science practices as discrete pieces of knowledge. Progression in learning was generally thought of as knowing more or providing more complete and correct responses. Similarly, practices were intentionally assessed in a way that minimized specific content knowledge demands—assessments were more likely to ask for definitions than for actual use of the practice. Assessment developers took this approach in part to be sure they were obtaining accurate measures of clearly definable constructs.[2] However, although understanding the language and termi-

---

[2]"Construct" is generally used to refer to concepts or ideas that cannot be directly observed, such as "liberty." In the context of educational measurement, the word is used more specifically to refer to a particular body of content (knowledge, understanding, or skills) that an assessment

nology of science is fundamental and factual knowledge is very important, tasks that demand only declarative knowledge about practices or isolated facts would be insufficient to measure performance expectations in the NGSS.

As we note in Chapter 3, the performance expectations provide a start in defining the claim or inference that is to be made about student proficiency. However, it is also important to determine the observations (the forms of evidence in student work) that are needed to support the claims, and then to develop tasks or situations that will elicit the needed evidence. The task development approaches described in Chapter 3 are commonly used for developing external tests, but they can also be useful in guiding the design of classroom assessments. Considering the intended inference, or claim, about student learning will help curriculum developers and classroom assessment designers ensure that the tasks elicit the needed evidence.

As we note in Chapter 2, assessment tasks aligned with the NGSS performance expectations will need to have multiple components—that is, be composed of more than one kind of activity or question. They will need to include opportunities for students to engage in practices as a means to demonstrate their capacity to apply them. For example, a task designed to elicit evidence that a student can develop and use models to support explanations about structure-function relationships in the context of a core idea will need to have several components. It may require that students articulate a claim about selected structure-function relationships, develop or describe a model that supports the claim, and provide a justification that links evidence to the claim (such as an explanation of an observed phenomenon described by the model). A multicomponent task may include some short-answer questions, possibly some carefully designed selected-response questions, and some extended-response elements that require students to demonstrate their understandings (such as tasks in which students design an investigation or explain a pattern of data). For the purpose of making an appraisal of student learning, no single piece of evidence is likely to be sufficient; rather, the pattern of evidence across multiple components can provide a sufficient indicator of student understanding.

---

is to measure. It can be used to refer to a very specific aspect of tested content (e.g., the water cycle) or a much broader area (e.g., mathematics).

Making Connections

The NGSS emphasize the importance of the connections among scientific concepts. Thus, the NGSS performance expectations for one disciplinary core idea may be connected to performance expectations for other core ideas, both within the same domain or in other domains, in multiple ways: one core idea may be a prerequisite for understanding another, or a task may be linked to more than one performance expectation and thus involve more than one practice in the context of a given core idea. NGSS-aligned tasks will need to be constructed so that they provide information about how well students make these connections. For example, a task that focused only on students' knowledge of a particular model would be less revealing than one that probed students' understanding of the kinds of questions and investigations that motivated the development of the model. Example 1, "What Is Going on Inside Me?" (in Chapter 2), shows how a single assessment task can be designed to yield evidence related to multiple performance expectations, such as applying physical science concepts in a life science context. Tasks that do not address these connections will not fully capture or adequately support three-dimensional science learning.

Learning as a Progression

The framework and the NGSS address the process of learning science. They make clear that students should be encouraged to take an investigative stance toward their own and others' ideas, to be open about what they are struggling to understand, and to recognize that struggle as part of the way science is done, as well as part of their own learning process. Thus, revealing students' emerging capabilities with science practices and their partially correct or incomplete understandings of core ideas is an important function of classroom assessment. The framework and the NGSS also postulate that students will develop disciplinary understandings by engaging in practices that help them to question and explain the functioning of natural and designed systems. Although learning is an ongoing process for both scientists and students, students are emerging practitioners of science, not scientists, and their ways of acting and reasoning differ from those of scientists in important ways. The framework discusses the importance of seeing learning as a trajectory in which students gradually progress in the course of a unit or a year, and across the whole K-12 span, and organizing instruction accordingly.

The first example in this chapter, "Measuring Silkworms" (also discussed in Chapter 3), illustrates how this idea works in an assessment that is embedded in a larger instructional unit. As they begin the task, students are not competent data analysts. They are unaware of how displays can convey ideas or of professional conventions for display and the rationale for these conventions. In designing their own displays, students begin to develop an understanding of the value of these conventions. Their partial and incomplete understandings of data visualization have to be explicitly identified so teachers can help them develop a more general understanding. Teachers help students learn about how different mathematical practices, such as ordering and counting data, influence the shapes the data take in models. The students come to understand how the shapes of the data support inferences about population growth.

Thus, as discussed in Chapter 2, uncovering students' incomplete forms of practice and understanding is critical: NGSS-aligned assessments will need to clearly define the forms of evidence associated with beginning, intermediate, and sophisticated levels of knowledge and practice expected for a particular instructional sequence. A key goal of classroom assessments is to help teachers and students understand what has been learned and what areas will require further attention. NGSS-aligned assessments will also need to identify likely misunderstandings, productive ideas of students that can be built upon, and interim goals for learning.

The NGSS performance expectations are general: they do not specify the kinds of intermediate understandings of disciplinary core ideas students may express during instruction nor do they help teachers interpret students' emerging capabilities with science practices or their partially correct or incomplete understanding. To teach toward the NGSS performance expectations, teachers will need a sense of the likely progression at a more micro level, to answer such questions as:

- For this unit, where are the students expected to start, and where should they arrive?
- What typical intermediate understandings emerge along this learning path?
- What common logical errors or alternative conceptions present barriers to the desired learning or resources for beginning instruction?
- What new aspects of a practice need to be developed in the context of this unit?

Classroom assessment probes will need to be designed to generate enough evidence about students' understandings so that their locations on the intended pathway can be reliably determined, and it is clear what next steps (instructional activities) are needed for them to continue to progress. As we note in Chapter 2, only a limited amount of research is available to support detailed learning progressions: assessment developers and others who have been applying this approach have used a combination of research and practical experience to support depictions of learning trajectories.

SIX EXAMPLES

We have identified six example tasks and task sets that illustrate the elements needed to assess the development of three-dimensional science learning. As noted in Chapter 1, they all predate the publication of the NGSS. However, the constructs being measured by each of these examples are similar to those found in the NGSS performance expectations. Each example was designed to provide evidence of students' capabilities in using one or more practices as they attempt to reach and present conclusions about one or more core ideas: that is, all of them assess three-dimensional learning. Table 1-1 shows the NGSS disciplinary core ideas, practices, and crosscutting ideas that are closest to the assessment targets for all of the examples in the report.[3]

We emphasize that there are many possible designs for activities or tasks that assess three-dimensional science learning—these six examples are only a sampling of the possible range. They demonstrate a variety of approaches, but they share some common attributes. All of them require students to use some aspects of one or more science and engineering practices in the course of demonstrating and defending their understanding of aspects of a disciplinary core idea. Each of them also includes multiple components, such as asking students to engage in an activity, to work independently on a modeling or other task, and to discuss their thinking or defend their argument.

These examples also show how one can use classroom work products and discussions as formative assessment opportunities. In addition, several of the examples include summative assessments. In each case, the evidence produced provides teachers with information about students' thinking and their developing understanding that would be useful for guiding next steps in instruction. Moreover, the time students spend in doing and reflecting on these tasks should

---

[3]The particular combinations in the examples may not be the same as NGSS examples at that grade level, but each of these examples of classroom assessment involves integrated knowledge of the same general type as the NGSS performance expectations. However, because they predate the NGSS and its emphasis on crosscutting concepts, only a few of these examples include reference to a crosscutting concept, and none of them attempts to assess student understanding of, or disposition to invoke, such concepts.

be seen as an integral part of instruction, rather than as a stand-alone assessment task. We note that the example assessment tasks also produce a variety of products and scorable evidence. For some we include illustrations of typical student work, and for others we include a construct map or scoring rubric used to guide the data interpretation process. Both are needed to develop an effective scoring system.

Each example has been used in classrooms to gather information about particular core ideas and practices. The examples are drawn from different grade levels and assess knowledge related to different disciplinary core ideas. Evidence from their use documents that, with appropriate prior instruction, students can successfully carry out these kinds of tasks. We describe and illustrate each of these examples below and close the chapter with general reflections about the examples, as well as our overall conclusions and recommendations about classroom assessment.

Example 3: Measuring Silkworms

The committee chose this example because it illustrates several of the characteristics we argue an assessment aligned with the NGSS must have: in particular, it allows the teacher to place students along a defined learning trajectory (see Figure 3-13 in Chapter 3), while assessing both a disciplinary core idea and a crosscutting concept.[4] The assessment component is formative, in that it helps the teacher understand what students already understood about data display and to adjust the instruction accordingly. This example, in which 3rd-grade students investigated the growth of silkworm larvae, first assesses students' conceptions of how data can be represented visually and then engages them in conversations about what different representations of the data they had collected reveal. It is closely tied to instruction—the assessment is embedded in a set of classroom activities.

The silkworm scenario is designed so that students' responses to the tasks can be interpreted in reference to a trajectory of increasingly sophisticated forms of reasoning. A construct map displayed in Figure 3-13 shows developing conceptions of data display. Once the students collect their data (measure the silkworms) and produce their own ways of visually representing their findings, the teacher uses the data displays as the basis for a discussion that has several objectives.

---

[4]This example is also discussed in Chapter 3 in the context of using construct modeling for task design.

The teacher uses the construct map to identify data displays that demonstrate several levels on the trajectory. In a whole-class discussion, she invites students to consider what the different ways of displaying the data "show and hide" about the data and how they do so. During this conversation, the students begin to appreciate the basis for conventions about display.[5] For example, in their initial attempt at representing the data they have collected, many of the students draw icons to resemble the organisms that are not of uniform size (see Figure 3-14 in Chapter 3). The mismatches between their icons and the actual relative lengths of the organisms become clear in the discussion. The teacher also invites students to consider how using mathematical ideas (related to ordering, counting, and intervals) helped them develop different shapes to represent the same data.

The teacher's focus on shape is an assessment of what is defined as the crosscutting concept of patterns in the framework and the NGSS. These activities also cultivate the students' capacity to think at a population level about the biological significance of the shapes, as they realize what the different representations of the measurements they have taken can tell them. Some of the student displays make a bell-like shape more evident, which inspires further questions and considerations in the whole-class discussion (see Figure 3-15 in Chapter 3): students notice that the tails of the distribution are comparatively sparse, especially for the longer larvae, and wonder why. As noted in Chapter 3, they speculate about the possible reasons for the differences, which leads to a discussion and conclusions about competition for resources, which in turn leads them to consider not only individual silkworms, but the entire population of silkworms. Hence, this assessment provides students with opportunities for learning about representations, while also providing the teacher with information about their understanding of a crosscutting concept (pattern) and disciplinary core concepts (population-level descriptions of variability and the mechanisms that produce it).

Example 4: Behavior of Air

The committee chose this example to show the use of classroom discourse to assess student understanding. The exercise is designed to focus students' attention on a particular concept: the teacher uses class discussion of the students' models of air particles to identify misunderstandings and then support students in collaboratively resolving them. This task assesses both students' understanding of the concept and their proficiency with the practices of modeling and developing oral

---

[5]This is a form of meta-representational competence; see diSessa (2004).

arguments about what they have observed. This assessment is used formatively and is closely tied to classroom instruction.

Classroom discussions can be a critical component of formative assessment. They provide a way for students to engage in scientific practices and for teachers to instantly monitor what the students do and do not understand. This example, from a unit for middle school students on the particle nature of matter, illustrates how a teacher can use discussions to assess students' progress and determine instructional next steps.[6]

In this example, 6th-grade students are asked to develop a model to explain the behavior of air. The activity leads them to an investigation of phase change and the nature of air. The example is from a single class period in a unit devoted to developing a conceptual model of a gas as an assemblage of moving particles with space between them; it consists of a structured task and a discussion guided by the teacher (Krajcik et al., 2013; Krajcik and Merritt, 2012). The teacher is aware of an area of potential difficulty for students, namely, a lack of understanding that there is empty space between the molecules of air. She uses group-developed models and student discussion of them as a probe to evaluate whether this understanding has been reached or needs further development.

When students come to this activity in the course of the unit, they have already reached consensus on several important ideas they can use in constructing their models. They have defined matter as anything that takes up space and has mass. They have concluded that gases—including air—are matter. They have determined through investigation that more air can be added to a container even when it already seems full and that air can be subtracted from a container without changing its size. They are thus left with questions about how more matter can be forced into a space that already seems to be full and what happens to matter when it spreads out to occupy more space. The students have learned from earlier teacher-led class discussions that simply stating that the gas changes "density" is not sufficient, since it only names the phenomenon—it does not indicate what actually makes it possible for differing amounts of gas to expand or contract to occupy the same space.

In this activity, students are given a syringe and asked to gradually pull the plunger in and out of it to explore the air pressure. They notice the pressure

---

[6]This example was drawn from research conducted on classroom enactments of the IQWST curriculum materials (Krajcik et al., 2008; Shwartz et al., 2008). In field trials of IQWST, a diverse group of students responded to the task described in this example: 43% were white/Asian and 57% were non-Asian/minority; and 4% were English learners (Banilower et al., 2010).

against their fingers when pushing in and the resistance as they pull the plunger out. They find that little or no air escapes when they manipulate the plunger. They are asked to work in small groups to develop a model to explain what happens to the air so that the same amount of it can occupy the syringe regardless of the volume of space available. The groups are asked to provide models of the air with the syringe in three positions: see Figure 4-1. This modeling activity itself is not used as a formal assessment task; rather, it is the class discussion, in which students compare their models, that allows the teacher to diagnose the students' understanding. That is, the assessment, which is intended to be formative, is conducted through the teacher's probing of students' understandings through classroom discussion.

Figure 4-2 shows the first models produced by five groups of students to depict the air in the syringe in its first position. The teacher asks the class to discuss the different models and to try to reach consensus on how to model the behavior of air to explain their observations. The class has agreed that there should be "air particles" (shown in each of their models as dark dots) and that the particles are moving (shown in some models by the arrows attached to the dots).

Most of their models are consistent in representing air as a mixture of different kinds of matter, including air, odor, dust, and "other particles." What is not consistent in their models is what is represented as *between* the particles: groups 1 and 2 show "wind" as the force moving the air particles; groups 3, 4, and 5 appear to show empty space between the particles. Exactly what, if anything, is in between the air particles emerges as a point of contention as the students discuss their models. After the class agrees that the consensus model should include air particles shown with arrows to demonstrate that the particles "are coming out in different directions," the teacher draws several particles with arrows and asks what to put next into the model. The actual classroom discussion is shown in Box 4-2.

The discussion shows how students engage in several scientific and engineering practices as they construct and defend their understanding about a disciplinary core idea. In this case, the key disciplinary idea is that there must be empty space between moving particles, which allows them to move, either to become more densely packed or to spread apart. The teacher can assess the way the students have drawn their models, which reveals that their understanding is not complete. They have agreed that all matter, including gas, is made of particles that are moving, but many of the students do not understand what is in between these moving particles. Several students indicate that they think there is air between the air par-

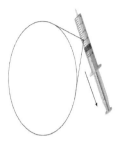

Model 1

Model 2

Model 3

FIGURE 4-1 Models for air in a syringe in three situations for Example 4, "Behavior of Air."
SOURCE: Krajcik et al. (2013). Reprinted with permission from Sangari Active Science.

ticles, since "air is everywhere," and some assert that the particles are all touching. Other students disagree that there can be air between the particles or that air particles are touching, although they do not yet articulate an argument for empty space between the particles, an idea that students begin to understand more clearly in subsequent lessons. Drawing on her observations, the teacher asks questions

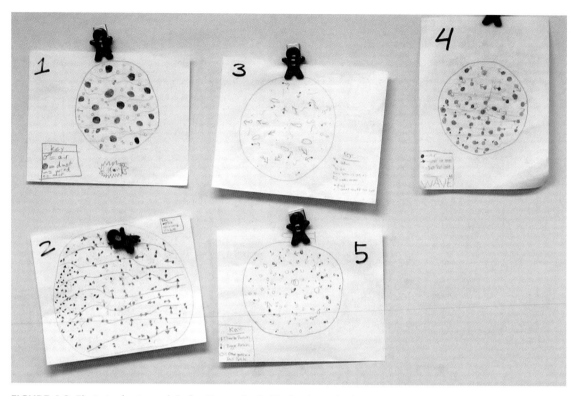

FIGURE 4-2 First student models for Example 4, "Behavior of Air."
SOURCE: Reiser et al. (2013). Copyright by the author; used with permission.

and gives comments that prompt the students to realize that they do not yet agree on the question of what is between the particles. The teacher then uses this observation to make instructional decisions. She follows up on one student's critique of the proposed addition to the consensus model to focus the students on their disagreement and then sends the class back into their groups to resolve the question.

In this example, the students' argument about the models plays two roles: it is an opportunity for students to defend or challenge their existing ideas, and it is an opportunity for the teacher to observe what the students are thinking and to decide that she needs to pursue the issue of what is between the particles of air. It is important to note that the teacher does not simply bring up this question, but instead uses the disagreement that emerges from the discussion as the basis for the question. (Later interviews with the teacher reveal that she had in fact anticipated that the empty space between particles would come up and was prepared to take advantage of that opportunity.) The discussion thus provides insights into students' thinking beyond their written (and drawn) responses to a task. The models themselves provide a context in which the students can clarify their thinking and refine their models in response to the critiques, to make more explicit claims to explain what they have observed. Thus, this activity focuses their attention on key explanatory issues (Reiser, 2004).

This example also illustrates the importance of engaging students in practices to help them develop understanding of disciplinary core ideas while also giving teachers information to guide instruction. In this case, the teacher's active probing of students' ideas demonstrates the way that formative assessment strategies can be effectively used as a part of instruction. The discussion of the models not only reveals the students' understanding about the phenomenon, but also allows the teacher to evaluate progress, uncover problematic issues, and help students construct and refine their models.

Example 5:   Movement of Water

The committee chose this example to show how a teacher can monitor developing understanding in the course of a lesson. "Clicker technology"[7] is used to obtain individual student responses that inform teachers of what the students have learned from an activity and which are then the basis for structuring small-group discussions that address misunderstandings. This task assesses both understanding of a concept as it develops in the course of a lesson and students' discussion skills. The assessments are used formatively and are closely tied to classroom instruction.

In the previous example (Example 4), the teacher orchestrates a discussion in which students present alternative points of view and then come to consensus about a disciplinary core idea through the practice of argumentation. However, many teachers may find it challenging to track students' thinking while also promoting the development of understanding for the whole class. The example on the movement of air was developed as part of a program for helping teachers learn to lead students in "assessment conversations" (Duschl and Gitomer, 1997).[8] In the

---

[7]Clicker technology, also known as classroom response systems, allows students to use handheld clickers to respond to questions from a teacher. The responses are gathered by a central receiver and immediately tallied for the teacher—or the whole class—to see.

[8]This example is taken from the Contingent Pedagogies Project, which provides formative assessment tools for middle schools and supports teachers in integrating assessment activities into discussions for both small groups and entire classes. Of the students who responded to the task, 46 percent were Latino. For more information, see http://contingentpedagogies.org [October 2013].

STUDENT-TEACHER DIALOGUE

Haley's objection: air is everywhere

Ms. B: OK. Now what?

S: Just draw like little. . . .

Haley: I think you should color the whole circle in, because dust . . . I mean air is everywhere, so. . . .

Miles: The whole circle?

Ms. B: So, I color the whole thing in.

Haley: Yeah.

Ms. B: So, if I do one like that, because I haven't seen one up here yet. If I color this whole thing in. . . .

[Ms. B colors in the whole region completely to show the air as Haley suggests.]

Michael: Then how would you show that . . . ?

Ms. B: Then ask . . . ask Haley some questions.

Students: How could that be? How would you show that?

Ms. B: Haley, people have some questions for you.

Some students object to Haley's proposal:

Frank: How would you show air?

Haley: Air is everywhere, so the air would be everything.

Ss: Yeah.

Alyssa: But then, how would you show the other molecules? I mean, you said air is everything, but then how would you show the other . . .?

Ss: Yeah, because . . . [Multiple students talking]

Haley: What? I didn't hear your question.

Alyssa: Um, I said if . . . You said air is everywhere, right?

Haley: Yeah. . . . so, that's why you wanted to color it in. But there's also other particles other than air, like dust and etc. and odors and things like that, so, how would you show that?

Miles: How are we going to put in the particles?

Ms. B: Haley, can you answer her?

Haley: No.

Ms. B: Why?

Haley: I don't know.

Other student: Because there is no way.

Ms. B: Why can't you answer?

Haley: What? I don't know.

Ms. B: Is what she's saying making sense?

Haley: Yeah.

Ms. B: What is it that you're thinking about?

Haley: Um . . . that maybe you should take . . . like, erase some of it to show the odors and stuff.

Addison: No, wait, wait!

Ms. B: All right, call on somebody else.

Addison proposes a compromise, and Ms. B pushes for clarification

Addison: Um, I have an idea. Like since air is everywhere, you might be able to like use a different colored marker and put like, um, the other molecules in there, so you're able to show that those are in there and then air is also everywhere.

Jerome: Yeah. I was gonna say that, or you could like erase it. If you make it all dark, you can just erase it and all of them will be.

Frank: Just erase some parts of the, uh . . . yeah, yeah, just to show there's something in between it.

Ms. B: And what's in between it?

Ss: The dust and the particles. Air particles. Other odors.

Miles: That's like the same thing over there.

Alyssa: No, the colors are switched.

Ms. B: Same thing over where?

Alyssa: The big one, the consensus.

Ms. B: On this one?

Alyssa: Yeah.

Ms. B: Well, what she's saying is that I should have black dots every which way, like that. [Ms. B draws the air particles touching one another in another representation, not in the consensus model, since it is Haley's idea.]

Students: No what? Yeah.

Ms. B: Right?

Students: No. Sort of. Yep.

Ms. B: OK. Talk to your partners. Is this what we want? [pointing to the air particles touching one another in the diagram]

Students discuss in groups whether air particles are touching or not, and what is between the particles if anything.

SOURCE: Reiser et al. (2013). Copyright by the author; used with permission.

task, middle school students engage in argumentation about disciplinary core ideas in earth science. As with the previous example, the formative assessment activity is more than just the initial question posed to students; it also includes the discussion that follows from student responses to it and teachers' decisions about what to do next, after she brings the discussion to a close.

In this activity, which also takes place in a single class session, the teacher structures a conversation about how the movement of water affects the deposition of surface and subsurface materials. The activity involves disciplinary core ideas (similar to Earth's systems in the NGSS) and engages students in practices, including modeling and constructing examples. It also requires students to reason about models of geosphere-hydrosphere interactions, which is an example of the cross-cutting concept pertaining to systems and system models.[9]

Teachers use classroom clicker technology to pose multiple-choice questions that are carefully designed to elicit students' ideas related to the movement of water. These questions have been tested in classrooms, and the response choices reflect common student ideas, including those that are especially problematic. In the course of both small-group and whole-class discussions, students construct and challenge possible explanations of the process of deposition. If students have difficulty in developing explanations, teachers can guide students to activities designed to improve their understanding, such as interpreting models of the deposition of surface and subsurface materials.

When students begin this activity, they will just have completed a set of investigations of weathering, erosion, and deposition that are part of a curriculum on investigating Earth systems.[10] Students will have had the opportunity to build physical models of these phenomena and frame hypotheses about how water will move sediment using stream tables.[11] The teacher begins the formative assessment activity by projecting on a screen a question about the process of deposition designed to check students' understanding of the activities they have completed: see Figure 4-3 for a sample question. Students select their answers using clickers.

---

[9]The specific NGSS core idea addressed is similar to MS-ESS2.C: "How do the properties and movement of water shape Earth's surface and affect its systems?" The closest NGSS performance expectation is MS-ESS2-c: "Construct an explanation based on evidence for how geoscience processes have changed Earth's surface at varying time and spatial scales."

[10]This curriculum, for middle school students, was developed by the American Geosciences Institute. For more information, see http://www.agiweb.org/education/ies [July 2013].

[11]Stream tables are models of stream flows set up in large boxes filled with sedimentary material and tilted so that water can flow through.

FIGURE 4-3 Sample question for Example 5, "Movement of Water."
SOURCE: NASA/GSFC/JPL/LaRC, MISR Science Team (2013) and Los Angeles County Museum of Art (2013).

Pairs or small groups of students then discuss their reasoning and offer explanations for their choices to the whole class. Teachers help students begin the small-group discussions by asking why someone might select A, B, or C, implying that any of them could be a reasonable response. Teachers press students for their reasoning and invite them to compare their own reasoning to that of others, using specific discussion strategies (see Michaels and O'Connor, 2011; National Research Council, 2007). After discussing their reasoning, students again vote, using their clickers. In this example, the student responses recorded using the clicker technology are scorable. A separate set of assessments (not discussed here) produces scores to evaluate the efficacy of the project as a whole.

The program materials include a set of "contingent activities" for teachers to use if students have difficulty meeting a performance expectation related to an investigation. Teachers use students' responses to decide which contingent activities are needed, and thus they use the activity as an informal formative assessment. In these activities, students might be asked to interpret models, construct explanations, and make predictions using those models as a way to deepen their understanding of Earth systems. In this example about the movement of air, students who are having difficulty understanding can view an animation of deposition and then make a prediction about a pattern they might expect to find at the mouth of a river where sediment is being deposited.

The aim of this kind of assessment activity is to guide teachers in using assessment techniques to improve student learning outcomes.[12] The techniques used in this example demonstrate a means of rapidly assessing how well students have mastered a complex combination of practices and concepts in the midst of a lesson, which allows teachers to immediately address areas students do not understand well. The contingent activities that provide alternative ways for students to master the core ideas (by engaging in particular practices) are an integral component of the formative assessment process.

Example 6:   Biodiversity in the Schoolyard

The committee chose this example to show the use of multiple interrelated tasks to assess a disciplinary core idea, biodiversity, with multiple science practices. As part of an extended unit, students complete four assessment tasks. The first three serve formative purposes and are designed to function close to instruction, informing the teacher about how well students have learned key concepts and mastered practices. The last assessment task serves a summative purpose, as an end-of-unit test, and is an example of a proximal assessment. The tasks address concepts related to biodiversity and science practices in an integrated fashion.

This set of four assessment tasks was designed to provide evidence of 5th-grade students' developing proficiency with a body of knowledge that blends a disciplinary core idea (biodiversity; LS4 in the NGSS; see Box 2-1 in Chapter 2) and a crosscutting concept (patterns) with three different practices: planning and carrying out investigations, analyzing and interpreting data, and constructing explanations (see Songer et al., 2009; Gotwals and Songer, 2013). These tasks, developed by researchers as part of an examination of the development of complex reasoning, are intended for use in an extended unit of study.[13]

---

[12]A quasi-experimental study compared the learning gains for students in classes that used the approach of the Contingent Pedagogies Project with gains for students in other classes in the same school district that used the same curriculum but not that approach. The students whose teachers used the Contingent Pedagogies Project demonstrated greater proficiency in earth science objectives than did students in classrooms in which teachers only had access to the regular curriculum materials (Penuel et al., 2012).

[13]The tasks were given to a sample of 6th-grade students in the Detroit Public School System, the majority of whom were racial/ethnic minority students (for details, see Songer et al., 2009).

Formative Assessment Tasks

Task 1: Collect data on the number of animals (abundance) and the number of different species (richness) in schoolyard zones.

Instructions: Once you have formed your team, your teacher will assign your team to a zone in the schoolyard. Your job is to go outside and spend approximately 40 minutes observing and recording all of the animals and signs of animals that you see in your schoolyard zone during that time. Use the BioKIDS application on your iPod to collect and record all your data and observations.

In responding to this task, students use an Apple iPod to record their information. The data from each iPod is uploaded and combined into a spreadsheet that contains all of the students' data; see Figure 4-4. Teachers use data from individual groups or from the whole class as assessment information to provide formative information about students' abilities to collect and record data for use in the other tasks.

Task 2: Create bar graphs that illustrate patterns in abundance and richness data from each of the schoolyard zones.

Task 2 assesses students' ability to construct and interpret graphs of the data they have collected (an important element of the NGSS practice "analyzing and interpreting data"). The exact instructions for Task 2 appear in Figure 4-5. Teachers use the graphs the students create for formative purposes, for making decisions about further instruction students may need. For example, if students are weak on the practices, the teacher may decide to help them with drawing accurate bars or the appropriate labeling of axes. Or if the students are weak on understanding of the core idea, the teacher might review the concepts of species abundance or species richness.

Task 3: Construct an explanation to support your answer to the question: Which zone of the schoolyard has the greatest biodiversity?

Before undertaking this task, students have completed an activity that helped them understand a definition of biodiversity: "An area is considered biodiverse if it has both a high animal abundance and high species richness." The students were also given hints (reminders) that there are three key parts of an explanation: a claim, more than one piece of evidence, and reasoning. The students are also

Animal Name	Zone A	Zone C	Zone E	Microhabitat	Total
Earthworms	2	0	2	– In dirt	4
Ants	2	229	75	– On something hard – On grass	306
Other insects	0	0	2	– On grass – Other microhabitat	2
Unknown beetle	0	3	0	– On plant	3
Unknown insect	0	2	0	– On dirt	2
Other leggy invertebrates	1	0	0	– In dirt	1
American robin	6	1	3	– On tree – In the sky	10
Black tern	200	0	0	– On plant – On something hard	200
House sparrow	0	0	1	– On tree	1
Mourning dove	3	0	0	– On tree	3
Unknown bird	7	5	2	– On tree – In the sky	14
Other birds	0	1	2	– In water – On grass	3
E. fox squirrel	1	1	0	– On something hard	2
Human	10	21	1	– On grass – On something hard – Other microhabitat	32
Other mammal	3	0	16	– Other microhabitat – On something hard	19
Red squirrel	2	0	0	– On tree	2
Number of Animals (Abundance)	237	263	104		604
Number of Kinds of Animals (Richness)	11	8	9		28

FIGURE 4-4 Class summary of animal observations in the schoolyard, organized by region (schoolyard zones), for Example 6, "Biodiversity in the Schoolyard."

Developing Assessments for the Next Generation Science Standards

Instructions:

1. Use your zone summary to make a bar chart of your **abundance** data.

Please remember to label your axes.

Our Abundance Data

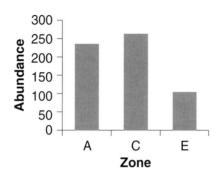

Result: According to the bar chart above, zone **C** has the highest abundance.

2. Use your zone summary to make a bar chart of your **richness** data.

Please remember to label your axes.

Our Richness Data

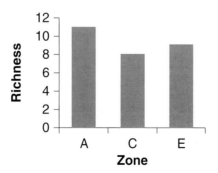

FIGURE 4-5 Instructions for Task 2 for Example 6, "Biodiversity in the Schoolyard." NOTE: See text for discussion.

given the definitions of relevant terms. This task allows the teacher to see how well students have understood the concept and can support their ideas about it. Instructions for Task 3 and student answers are shown in Box 4-3.

Summative Assessment Task

Task 4: Construct an explanation to support an answer to the question: Which zone of the schoolyard has the greatest biodiversity?

For the end-of-unit assessment, the task presents students with excerpts from a class data collection summary, shown in Table 4-1, and asks them to construct an explanation, as they did in Task 3. The difference is that in Task 4, the hints are removed: at the end of the unit, they are expected to show that they understand what constitutes a full explanation without a reminder. The task and coding rubric used for Task 4 are shown in Box 4-4.

The Set of Tasks

This set of tasks illustrates two points. First, using tasks to assess several practices in the context of a core idea together with a crosscutting concept can provide a wider range of information about students' progression than would tasks that focused on only one practice. Second, classroom assessment tasks in which core ideas, crosscutting concepts, and practices are integrated can be used for both formative and summative purposes. Table 4-2 shows the core idea, crosscutting concept, practices, assessment purposes, and performance expectation targets for assessment for each of the tasks. Each of these four tasks was designed to provide information about a single performance expectation related to the core idea, and each performance expectation focused on one of three practices. Figure 4-6 illustrates the way these elements fit together to identify the target for assessment of Tasks 3 and 4.

Second, the design of each task was determined by its purpose (formative or summative) and the point in the curriculum at which it was to be used. Assessment tasks may, by design, include more or less guidance for students, depending on the type of information they are intended to collect. Because learning is a process that occurs over time, a teacher might choose an assessment task with fewer guides (or scaffolds) for students as they progress through a curriculum to gather evidence of what students can demonstrate without assistance. Thus, the task developers offered a practice progression to illustrate the different levels of

BOX 4-3

INSTRUCTIONS AND SAMPLE STUDENT ANSWERS FOR TASK 3 IN EXAMPLE 6, "BIODIVERSITY IN THE SCHOOLYARD"

Instructions: Using what you have learned about biodiversity, the information from your class summary sheet, and your bar charts for abundance and richness, construct an explanation to answer the following scientific question:

Scientific Question: Which zone in the schoolyard has the highest biodiversity?
My Explanation [figure or text box?]

Make a CLAIM: Write a complete sentence that answers the scientific question.

Zone A has the greatest biodiversity.

Hint: Look at your abundance and richness data sheets carefully.

Give your REASONING: Write the scientific concept or definition that you thought about to make your claim.

Hint: Think about how biodiversity is related to abundance and richness.

Biodiversity is related to abundance and richness because it shows the two amounts in one word.

Give your EVIDENCE: Look at your data and find two pieces of evidence that help answer the scientific question.

Hint: Think about which zone has the highest abundance and richness.

1. Zone A has the most richness.

2. Zone A has a lot of abundance.

NOTES: Student responses are shown in italics. See text for discussion.

TABLE 4-1 Schoolyard Animal Data for Example 6 Summative Task, "Biodiversity in the Schoolyard"

Animal Name	Zone A	Zone B	Zone C	Total
Pillbugs	1	3	4	8
Ants	4	6	10	20
Robins	0	2	0	2
Squirrels	0	2	2	4
Pigeons	1	1	0	2
Animal abundance	6	14	16	36
Animal richness	3	5	3	5

guidance that tasks might include, depending on their purpose and the stage students will have reached in the curriculum when they undertake the tasks.

Box 4-5 shows a progression for the design of tasks that assess one example of three-dimensional learning: the practice of constructing explanations with one core idea and crosscutting concept. This progression design was based on studies that examined students' development of three-dimensional learning over time, which showed that students need less support in tackling assessment tasks as they progress in knowledge development (see, e.g., Songer et al., 2009).

Tasks 3 and 4, which target the same performance expectation but have different assessment purposes, illustrate this point. Task 3 was implemented midway through the curricular unit to provide formative information for the teacher on the kinds of three-dimensional learning students could demonstrate with the assistance of guides. Task 3 was classified as a Level 5 task (in terms of the progression shown in Box 4-5) and included two types of guides for the students (core idea guides in text boxes and practice guides that offer the definition of claim, evidence, and reasoning). Task 4 was classified as a Level 7 task because it did not provide students with any guides to the construction of explanations.

Example 7: Climate Change

The committee chose this flexible online assessment task to demonstrate how assessment can be customized to suit different purposes. It was designed to probe student understanding and to facilitate a teacher's review of responses. Computer software allows teachers to tailor online assessment tasks to their purpose and to the stage of learning that students have reached, by offering more or less supporting information

BOX 4-4

TASK AND CODING RUBRIC FOR TASK 4 IN EXAMPLE 6, "BIODIVERSITY IN THE SCHOOLYARD"

Write a scientific argument to support your answer for the following question.

Scientific Question: Which zone has the highest biodiversity?

Coding

4 points: Contains all parts of explanation (correct claim, 2 pieces of evidence, reasoning)
3 points: Contains correct claim and 2 pieces of evidence but incorrect or no reasoning
2 points: Contains correct claim + 1 piece correct evidence OR 2 pieces correct evidence and 1 piece incorrect evidence
1 point: Contains correct claim, but no evidence or incorrect evidence and incorrect or no reasoning

Correct Responses

Claim

Correct: Zone B has the highest biodiversity.

Evidence

1. Zone B has the highest animal richness.
2. Zone B has high animal abundance.

Reasoning

Explicit written statement that ties evidence to claim with a reasoning statement: that is, Zone B has the highest biodiversity because it has the highest animal richness and high animal abundance. *Biodiversity is a combination of both richness and abundance*, not just one or the other.

and guidance. The tasks may be used for both formative and summative purposes: they are designed to function close to instruction.

This online assessment task is part of a climate change curriculum for high school students. It targets the performance expectation that students use geoscience data and the results from global climate models to make evidence-based

TABLE 4-2 Characteristics of Tasks in Example 6, "Biodiversity in the Schoolyard"

Core Idea	Crosscutting Concepts	Practices	Purpose of Assessment	Target for Assessment: Performance Expectation
LS4.D Biodiversity and Humans	Patterns	Planning and carrying out investigations	Formative	Task 1. Collect data on the number of animals (abundance) and the number of different species (richness) in schoolyard zones.
		Analyzing and interpreting data	Formative	Task 2. Create bar graphs that illustrate patterns in abundance and richness data from each of the schoolyard zones.
		Constructing explanations	Formative	Task 3. Construct an explanation to support your answer to the question: Which zone of the schoolyard has the greatest biodiversity?
		Constructing explanations	Summative	Task 4. Construct an explanation to support your answer to the question: Which zone of the schoolyard has the greatest biodiversity?

forecasts of the impacts of climate change on organisms and ecosystems.[14] This example illustrates four potential benefits of online assessment tasks:

1. the capacity to present data from various external sources to students;
2. the capacity to make information about the quality and range of student responses continuously available to teachers so they can be used for formative purposes;
3. the possibility that tasks can be modified to provide more or less support, or scaffolding, depending on the point in the curriculum at which the task is being used; and

---

[14]This performance expectation is similar to two in the NGSS ones: HS-LS2-2 and HS-ESS3-5, which cover the scientific practices of analyzing and interpreting data and obtaining, evaluating, and communicating evidence.

Practice-Constructing Explanations (5th grade) = the use of evidence in constructing explanations that specify variables that describe and predict phenomena		Crosscutting (patterns) and Core Idea (biodiversity): Biodiversity describes the variety of species found in Earth's terrestrial and oceanic systems. The completeness of integrity of an ecosystem's biodiversity is often used as a measure of its health schoolyard zones.

Blended Knowledge Task 3 and 4 Construct

Construct an explanation to support your answer to the question, which zone of the schoolyard has the greatest biodiversity?

FIGURE 4-6 Combining practice, crosscutting concept, and core idea to form a blended learning performance expectation, assessed in Tasks 3 and 4, for Example 6, "Biodiversity in the Schoolyard."

4. the possibility that the tasks can be modified to be more or less active depending on teachers' or students' preferences.

In the instruction that takes place prior to this task, students will have selected a focal species in a particular ecosystem and studied its needs and how it is distributed in the ecosystem. They will also have become familiar with a set of model-based climate projections, called Future 1, 2, and 3, that represent more and less severe climate change effects. Those projections are taken from the Intergovernmental Panel on Climate Change (IPCC) data predictions for the year 2100 (Intergovernmental Panel on Climate Change, 2007): see Figure 4-7. The materials provided online as part of the activity include

- global climate model information presented in a table showing three different IPCC climate change scenarios (shown in Figure 4-7);
- geosciences data in the form of a map of North America that illustrates the current and the predicted distribution of locations of optimal biotic and abiotic[15] conditions for a species, as predicted by IPCC Future 3 scenario: see Figure 4-8; and

---

[15]The biotic component of an environment consists of the living species that populate it, while the abiotic components are the nonliving influences such as geography, soil, water, and climate that are specific to the particular region.

PROGRESSION FOR MULTIDIMENSIONAL LEARNING TASK DESIGN

This progression covers constructing a claim with evidence and constructing explanations with and without guidance. The + and ++ symbols represent the number of guides provided in the task.

Level 7: Student is provided with a question and is asked to construct a scientific explanation (no guides).

Level 6+: Student is provided with a question and is asked to construct a scientific explanation (with core ideas guides only).

Level 5++: Student is provided with a question and is asked to construct a scientific explanation (with core ideas guides and guides defining claim, evidence and reasoning).

Level 4: Student is provided with a question and is asked to make a claim and back it with evidence (no guides).

Level 3+: Student is provided with a question and is asked to make a claim and back it with evidence (with core ideas guides only).

Level 2++: Student is provided with a question and is asked to make a claim and back it with evidence (with core ideas guides and guides defining claim and evidence).

Level 1: Student is provided with evidence and asked to choose appropriate claim OR student is provided with a claim and is asked to choose the appropriate evidence.

SOURCE: Adapted from Gotwals and Songer (2013).

- an online guide for students in the development of predictions, which prompts them as to what is needed and records their responses in a database that teachers and students can use. (The teacher can choose whether or not to allow students access to the pop-up text that describes what is meant by a claim or by evidence.)

The task asks students to make and support a prediction in answer to the question, "In Future 3, would climate change impact your focal species?" Students are asked to provide the following:

	Population growth rate	Energy use per person	Proportion clean energy	Total CO_2 emissions by 2100 (gigatons)
Future 1	Fast	Low	Low	1862
Future 2	Slow	High	High	1499
Future 3	Slow	Low	High	983

FIGURE 4-7 Three simplified Intergovernmental Panel on Climate Change (IPCC)-modeled future scenarios for the year 2100.
SOURCE: Adapted from Peters et al. (2012).

- a claim (the prediction) as to whether or not they believe the IPCC scenario information suggests that climate change will affect their chosen animal;
- reasoning that connects their prediction to the model-based evidence, such as noting that their species needs a particular prey to survive; and
- model-based evidence that is drawn from the information in the maps of model-based climate projections, such as whether or not the distribution of conditions needed by the animal and its food source in the future scenario will be significantly different from what it is at present.

Table 4-3 shows sample student responses that illustrate both correct responses and common errors. Students 1, 3, and 4 have made accurate predictions, and supplied reasoning and evidence; students 2, 5, and 6 demonstrate common errors, including insufficient evidence (student 2), inappropriate reasoning and evidence (student 5), and confusion between reasoning and evidence (student 6). Teachers can use this display to quickly see the range of responses in the class and use that information to make decisions about future instruction.

Example 8:   Ecosystems

The committee chose this example, drawn from the SimScientists project, to demonstrate the use of simulation-based modules designed to be embedded in a curriculum unit to provide both formative and summative assessment information. Middle school students use computer simulations to demonstrate their understanding of core ideas about ecosystem dynamics and the progress of their

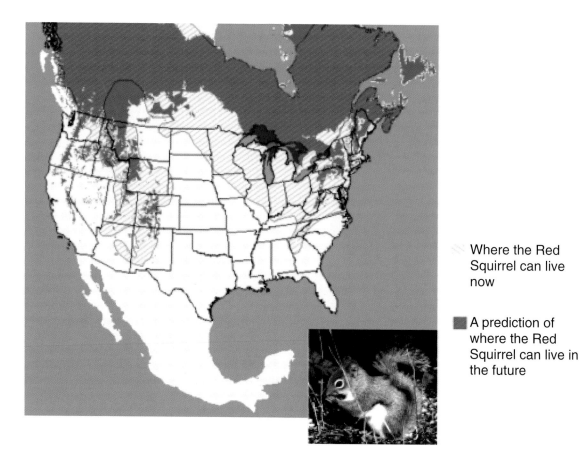

Where the Red Squirrel can live now

A prediction of where the Red Squirrel can live in the future

FIGURE 4-8 Current and predicted Future 3 distribution for the red squirrel for Example 7, "Climate Change."
SOURCE: Songer et al. (2013). Copyright by the author; used with permission.

thinking as they move from exploring ecosystem components to interactions of those components to the way systems behave. Thus, the simulations also address the crosscutting concept of systems. The assessment components function close to classroom instruction.

In this set of classroom modules, students use simulated, dynamic representations of particular ecosystems, such as a mountain lake or grasslands, to investigate features common to all ecosystems. The students investigate the roles of and relationships among species within habitats and the effects of these interactions on population levels (Quellmalz et al., 2009). Simulations of these environments can be used both to improve students' understanding of complex ecosystems and to

TABLE 4-3 Sample Student Responses in Example 7, "Climate Change"

Scientific Question: In Future 3, would climate change impact your focal species?

Student 1	Claim	Climate change will effect my focal species.
	Reasoning	The abiotic conditions will change, and the temperature will change; therefore, the habitat of my species will change.
	Evidence	The map shows it will move into the Western part, therefore the climate changed.
Student 2	Claim	Yes it will effect it, it will shorten the range.
	Reasoning	When the climate changes the focal species will have to move north because it won't be able to stand the warm weather.
	Evidence	The map.
Student 3	Claim	Yes, climate change would effect the red-backed salamander.
	Reasoning	Abiotic and biotic factors can cause the red-backed salamander to relocate, such as temperature, precipitation, and invasive species.
	Evidence	If the temperatures increase, the red-backed salamander would have to live farther north where temperatures are suitable for its survival.
Student 4	Claim	I think that climate change in Future 3 will not impact my focal species.
	Reasoning	Some abiotic features that could affect the focal species could be the climate, but it won't move the focal species from the location.
	Evidence	According to the distribution map for Future 3 the American Kestrel does not move from the location.
Student 5	Claim	No because my focal species is a bird and it can migrate to a warmer area but if the climate gets warm earlier then it will migrate earlier and it could affect it's normal time to migrate.
	Reasoning	The food they eat might not be out of hibernation or done growing in the area it migrates to.
	Evidence	It eats mice and mice hibernate and and so do voles and if the climate changes to a cold climate too early then their food will be hidden and they will have to migrate early.

NOTE: Both correct and incorrect responses are shown.

SOURCE: Songer et al. (2013). Copyright by the author; used with permission.

Model Level	Description	Disciplinary Core Ideas	Science Practices
Component	What are the components of the system and their rules of behavior?	Every ecosystem has a similar pattern of organization with respect to the roles (producers, consumers, and decomposers) that organisms play in the movement of energy and matter through the system. (NGSS: LS2.A — Interdependent relationships in Ecosystems)	Analyze and interpret data to provide evidence for phenomena.
Interaction	How do the the individual components interact?	Matter and energy flow through the ecosystem as individual organisms participate in feeding relationships within an ecosystem. (NGSS: LS2.B — Cycles of Matter and Energy Transfer in Ecosystems)	Develop a model to describe phenomena. Analyze and interpret data.
Emergent	What is the overall behavior or property of the system that results from many interactions following specific rules?	Interactions among organisms and among organisms and the ecosystem's nonliving features cause the populations of the different organisms to change over time. (NGSS: LS2.C — Ecosystem Dynamics, Functioning, and Resilience)	Use a model to plan and carry out investigations. Analyze and interpret data to provide evidence.

FIGURE 4-9 Ecosystems target model for Example 8, "Ecosystems."
SOURCE: SimScientists Calipers II project (2013). Reprinted with permission.

assess what they have learned. The simulated environments provide multiple representations of system models at different scales. They require students to apply core ideas about ecosystems and to carry out such practices as building and using models, planning and conducting investigations (by manipulating the system elements), and interpreting patterns.

Figure 4-9 shows a model of the characteristics of and changes in ecosystems as it would appear on the screen. The model would be very difficult for students to observe or investigate using printed curriculum materials.[16] For example, Figure 4-10 shows part of a simulated mountain lake environment. Students observe animations of the organisms' interactions and are then asked to draw a food web directly on the screen to represent a model of the flow of matter and energy in the ecosystem. If a student draws an arrow that links a food consumer to the wrong source of matter and energy, a feedback box coaches the student to observe again by reviewing the animation, thus providing formative feedback.

---

[16]These same features also make it difficult to display the full impact of the simulation in this report.

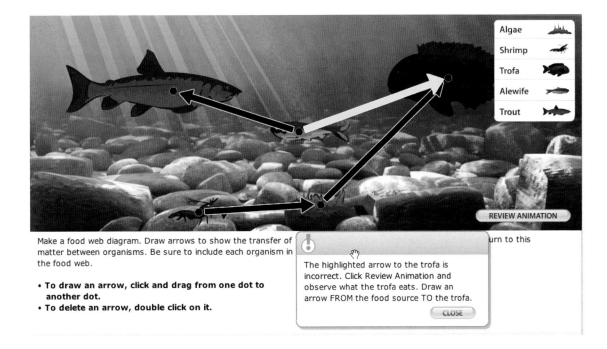

FIGURE 4-10 Screenshot of a curriculum-embedded assessment of student constructing a food web to model the flow of matter and energy in the ecosystem (with feedback and coaching); part of Example 8, "Ecosystems."
SOURCE: Quellmalz et al. (2012, fig. 2, p. 372). Reprinted with permission from John Wiley & Sons.

In the subsequent curriculum-embedded assessment, students investigate what happens to population levels when relative starting numbers of particular organisms are varied: see Figure 4-11. The interactive simulation allows students to conduct multiple trials to build, evaluate, and critique models of balanced ecosystems, interpret data, and draw conclusions. If the purpose of the assessment is formative, students can be given feedback and a graduated sequence of coaching by the program. Figure 4-11 shows a feedback box for this set of activities, which not only notifies the student that an error has occurred but also prompts the student to analyze the population graphs and design a third trial that maintains the survival of the organisms. As part of the assessment, students also complete tasks that ask them to construct descriptions, explanations, and conclusions. They are guided in assessing their own work by judging whether their response meets specified criteria, and then how well their response matches a sample one, as illustrated in Figure 4-12.

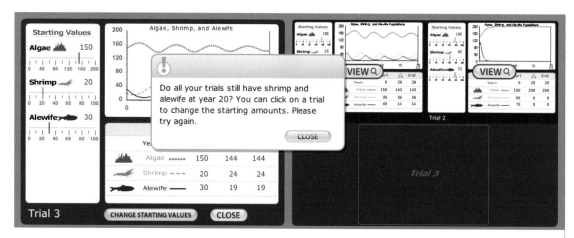

Can you do better than Dr. A? Design three trials so that both the shrimp and alewife populations survive for 20 years.

- **Use the sliders to change the starting numbers of shrimp and alewife.**
- **Click RUN to see what happens.**
- **When you have saved 3 trials in which shrimp and alewife survive for 20 years, click NEXT.**

FIGURE 4-11 Screenshot of a curriculum-embedded assessment of student using simulations to build balanced ecosystem population models (with feedback and coaching); part of Example 8, "Ecosystems."

SOURCE: SimScientists Calipers II project (2013). Reprinted with permission.

The SimScientists assessments are designed to provide feedback that addresses common student misconceptions about the ecosystem components, interactions that take place within them, or the way they behave, as well as errors in the use of science practices. The simulation generates reports to students about their progress toward goals for conceptual understanding and use of practices, and it also provides a variety of reporting options for teachers. Teachers can view progress reports for individual students as well as class-level reports (Quellmalz et al., 2012).

The SimScientists assessment system was also designed to collect summative assessment information after students complete a regular curriculum unit on ecosystems (which might have included the formative assessment modules described above). Figures 4-13 and 4-14 show tasks that are part of a benchmark assessment scenario in which students are asked to investigate ways to restore an Australian grasslands ecosystem—one that is novel to them—that has been affected by a significant fire. No feedback or coaching is provided. Students investigate the roles of

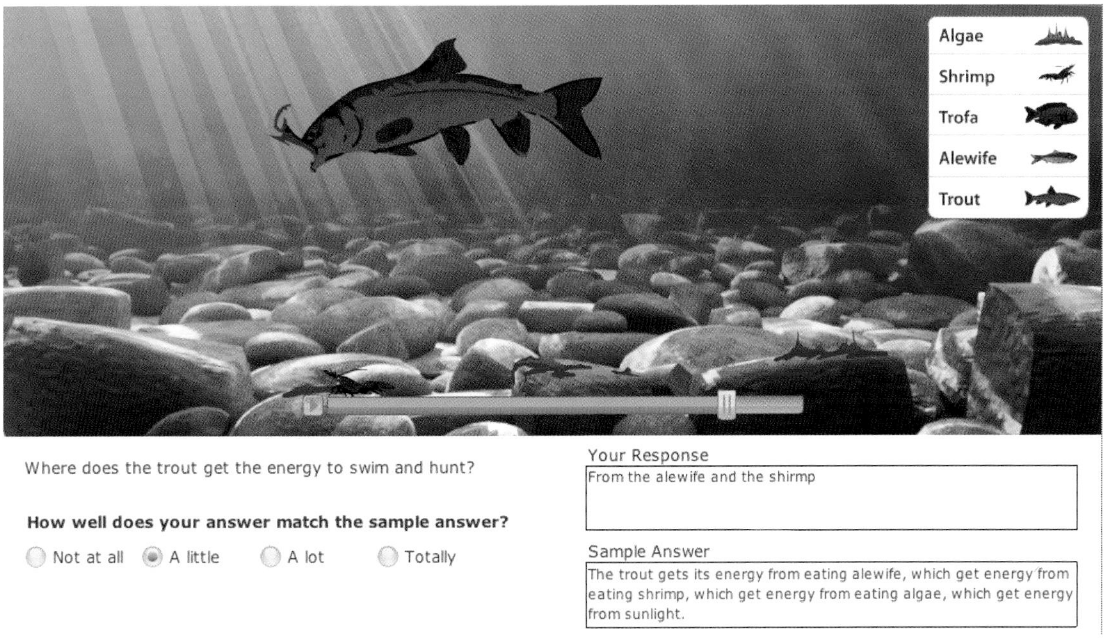

FIGURE 4-12 Screenshot of a curriculum-embedded assessment of student comparing his/her constructed response describing the mountain lake matter and energy flow model to a sample response; part of Example 8, "Ecosystems."
SOURCE: SimScientists Calipers II project (2013). Reprinted with permission.

and relationships among the animals, birds, insects, and grass by observing animations of their interactions. Students draw a food web representing a model of the flow of energy and matter throughout the ecosystem, based on the interactions they have observed. Students then use the simulation models to plan, conduct, interpret, explain, and critique investigations of what happens to population levels when numbers of particular organisms are varied. In a culminating task, students present their findings about the grasslands ecosystem.

These task examples from the SimScientists project illustrate ways that assessment tasks can take advantage of technology to represent generalizable, progressively more complex models of science systems, present challenging scientific reasoning tasks, provide individualized feedback, customize scaffolding, and promote self-assessment and metacognitive skills. Reports generated for teachers and students indicate the level of additional help students may need and classify students into groups for which tailored, follow-on, reflection activities are recommended (to be conducted during a subsequent class period).

Make a food web diagram. Draw arrows to show the transfer of matter between organisms.

Be sure to include each organism in the food web.

- **To draw an arrow, click and drag from one dot to another dot.**
- **To delete an arrow, double click on it.**

You can review the animation and then return to this diagram.

FIGURE 4-13 Screenshot of a benchmark summative assessment of a student constructing a food web to model the flow of matter and energy in the ecosystem (without feedback and coaching); part of Example 8, "Ecosystems."

SOURCE: SimScientists Calipers II project (2013). Reprinted with permission.

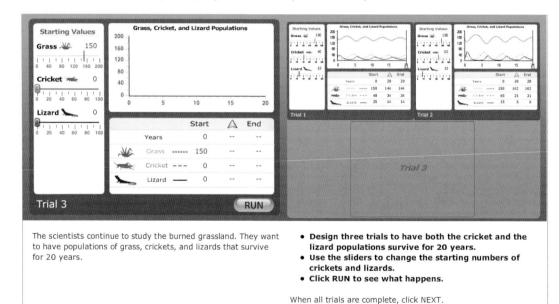

The scientists continue to study the burned grassland. They want to have populations of grass, crickets, and lizards that survive for 20 years.

- **Design three trials to have both the cricket and the lizard populations survive for 20 years.**
- **Use the sliders to change the starting numbers of crickets and lizards.**
- **Click RUN to see what happens.**

When all trials are complete, click NEXT.

FIGURE 4-14 Screenshot of a benchmark summative assessment of a student using simulations to build balanced ecosystem population models (without feedback and coaching); part of Example 8, "Ecosystems."

SOURCE: SimScientists Calipers II project (2013). Reprinted with permission.

These formative assessments also have an instructional purpose. They are designed to promote model-based reasoning about the common organization and behaviors of all ecosystems (see Figure 4-9) and to teach students how to transfer knowledge they gain about how one ecosystem functions to examples of new ecosystems (Buckley and Quellmalz, 2013).[17]

LESSONS FROM THE EXAMPLES

The six examples discussed above, as well as the one in Chapter 2, demonstrate characteristics we believe are needed to assess the learning called for in the NGSS and a range of approaches to using assessments constructively in the classroom to support such learning. A key goal of classroom assessment is to elicit and make visible students' ways of thinking and acting. The examples demonstrate that it is possible to design tasks and contexts in which teachers elicit student thinking about a disciplinary core idea or crosscutting concept by engaging them in a scientific practice. The examples involve activities designed to stimulate classroom conversations or to produce a range of artifacts (products) that provide information to teachers about students' current ways of thinking and acting, or both. This information can be used to adjust instruction or to evaluate learning that occurred during a specified time. Some of the examples involve formal scoring, while others are used by teachers to adjust their instructional activities without necessarily assigning student scores.

Types of Assessment Activities

In "What Is Going on Inside Me?" (Example 1 in Chapter 2), students produce a written evidence-based argument for an explanation of how animals get energy from food and defend that explanation orally in front of the class. In "Measuring Silkworms" (Example 3, above, and also discussed in Chapter 3), students produce representations of data and discuss what they do and do not reveal about the data. In "Behavior of Air" (Example 4, above), models developed by groups of students are the stimulus for class discussion and argumentation that the teacher uses to diagnose and highlight discrepancies in students' ideas. In "Movement of Water" (Example 5, above), multiple-choice questions that students answer using

---

[17]The system was designed using the evidence-centered design approach discussed in Chapter 3. Research on the assessments supports the idea that this approach could be a part of a coherent, balanced state science assessment system: see discussion in Chapter 6.

clickers are the stimulus for class discussion (assessment conversation). In each of these examples, students' writing and classroom discourse provide evidence that can be used in decisions about whether additional activities for learning might be needed, and, if so, what kinds of activities might be most productive. In many of these examples, listening to and engaging with other students as they discuss and defend their responses is a part of the learning process, as students work toward a classroom consensus explanation or a model based on the evidence they have collected. The classroom discussion itself in these cases is the basis for the formative assessment process.

We note that when assessments are designed to be used formatively, the goal is sometimes not to assign scores to individual students but rather to decide what further instruction is needed for groups of students or the class as a whole. Thus, instead of scoring rubrics, criteria or rubrics that can help guide instructional decisions may be used. (When the goal includes assessment of both individuals and groups, both types of scoring rubrics would be needed.) Teachers need support to learn to be intentional and deliberative about such decisions. In the examples shown, designers of curriculum and instruction have developed probes that address likely learning challenges, and teachers are supported in recognizing these challenges and in the use of the probes to seek evidence of what their students have learned and not learned, along some continuum.

"Ecosystems" (Example 8, above) is a computer-based system in which students use simulations both to learn and to demonstrate what they have learned about food webs. It includes tasks that are explicitly designed for assessment. Other tasks may not be sharply distinguished from ongoing classroom activities. The data collection tasks in "Biodiversity in the Schoolyard" (Example 6, above) are part of students' ongoing investigations, not separate from them, but they can provide evidence that can be used for formative purposes.

Similarly, in "Measuring Silkworms" (Example 3) students create displays as part of the learning process in order to answer questions about biological growth. Constructing these displays engages students in the practice of analyzing data, and their displays are also a source of evidence for teachers about students' proficiencies in reasoning about data aggregations; thus they can be used formatively. These forms of reasoning also become a topic of instructional conversations, so that students are encouraged to consider additional aspects of data representation, including tradeoffs about what different kinds of displays do and do not show about the same data. As students improve their capacity to visualize data, the data discussion then leads them to notice characteristics of organisms or populations that are otherwise not apparent. This interplay between learning a practice (data representation as an aspect of data analysis) and learning about a core idea (variation in a population), as well as a crosscutting concept (recognizing and interpreting patterns), provides an example of the power of three-dimensional learning, as well as an example of an assessment strategy.

Interpreting Results

A structured framework for interpreting evidence of student thinking is needed to make use of the task artifacts (products), which might include data displays, written explanations, or oral arguments. As we discuss in Chapter 3, interpretation of results is a core element of assessment, and it should be a part of the assessment design. An interpretive framework can help teachers and students themselves recognize how far they have progressed and identify intermediate stages of understanding and problematic ideas. "Measuring Silkworms" shows one such framework, a learning progression for data display developed jointly by researchers and teachers. "Behavior of Air" is similarly grounded in a learning progressions approach. "Movement of Water" presents an alternative example, using what is called a facets-based approach[18] to track the stages in a learning progression (discussed in Chapter 2)—that is, to identify ideas that are commonly held by students relative to a disciplinary core idea. Although these preconceptions are often labeled as misconceptions or problematic ideas, they are the base on which student learning must be built. Diagnosing students' preconceptions can help teachers identify the types of instruction needed to move students toward a more scientific conception of the topic.

What these examples have in common is that they allow teachers to group students into categories, which helps with the difficult task of making sense of many kinds of student thinking; they also provide tools for helping teachers decide what to do next. In "Movement of Water," for example, students' use of clickers

---

[18]In this approach, a facet is a piece of knowledge constructed by a learner in order to solve a problem or explain an event (diSessa and Minstrell, 1998). Facets that are related to one another can be organized into clusters, and the basis for grouping can either be an explanation or an interpretation of a physical situation or a disciplinary core idea (Minstrell and Kraus, 2005). Clusters comprise goal facets (which are often standards or disciplinary core ideas) and problematic facets (which are related to the disciplinary idea but which represent ways of reasoning about the idea that diverge from the goal facet). The facets perspective assumes that, in addition to problematic thinking, students also possess insights and understandings about the disciplinary core idea that can be deepened and revised through additional learning opportunities (Minstrell and van Zee, 2003).

to answer questions gives teachers initial feedback on the distribution of student ideas in the classroom. Depending on the prevalence of particular problematic ideas or forms of reasoning and their persistence in subsequent class discussion, teachers can choose to use a "contingent activity" that provides a different way of presenting a disciplinary core idea.

The interpretive framework for evaluating evidence has to be expressed with enough specificity to make it useful for helping teachers decide on next steps. The construct map for data display in "Measuring Silkworms" meets this requirement: a representation that articulated only the distinction between the lowest and highest levels of the construct map would be less useful. Learning progressions that articulate points of transition that take place across multiple years—rather than transitions that may occur in the course of a lesson or a unit—would be less useful for classroom decision making (although a single classroom may often include students who span such a range) (Alonzo and Gearhart, 2006).

Using Multiple Practices

The examples above involve tasks that cross different domains of science and cover multiple practices. "What Is Going on Inside Me?," for example, requires students to demonstrate their understanding of how chemical processes support biological processes. It asks students not only to apply the crosscutting concept of energy and matter conservation, but also to support their arguments with explicit evidence about the chemical mechanism involved. In "Measuring Silkworms" and "Biodiversity in the Schoolyard," students' responses to the different tasks can provide evidence of their understanding of the crosscutting concept of patterns. It is important to note, however, that "patterns" in each case has a different and particular disciplinary interpretation. In "Measuring Silkworms," students must recognize pattern in a display of data, in the form of the "shapes" the data can take, and begin to link ideas about growth and variation to these shapes. In contrast, in "Biodiversity in the Schoolyard," students need to recognize patterns in the distribution and numbers of organisms in order to use the data in constructing arguments.

Three of the examples—"Measuring Silkworms," "Biodiversity in the Schoolyard," and "Climate Change"—provide some classroom-level snapshots of emerging proficiency with aspects of the practices of analyzing and interpreting data and using mathematics and computational thinking. We note, though, that each of these practices has multiple aspects, so multiple tasks would be needed to provide a complete picture of students' capacity with each of them. Although assessment tasks can identify particular skills related to specific practices, evaluating students' disposition to engage in these practices without prompting likely requires some form of direct observation or assessment of the products of more open-ended student projects.[19]

In instruction, students engage in practices in interconnected ways that support their ongoing investigations of phenomena. Thus, students are likely to find that to address their questions, they will need to decide which sorts of data (including observational data) are needed; that is, they will need to design an investigation, collect those data, interpret the results; and construct explanations that relate their evidence to both claims and reasoning. It makes little sense for students to construct data displays in the absence of a question. And it is not possible to assess the adequacy of their displays without knowing what question they are pursuing. In the past, teachers might have tried to isolate the skill of graphing data as something to teach separately from disciplinary content, but the new science framework and the NGSS call for teachers to structure tasks and interpret evidence in a broad context of learning that integrates or connects multiple content ideas and treats scientific practices as interrelated. Similarly, assessment tasks designed to examine students' facility with a particular practice may require students to draw on other practices as they complete the task.

We stress in Chapter 2 that a key principle of the framework is that science education should connect to students' interests and experiences. Students are likely to bring diverse interests and experiences to the classroom from their families and cultural communities. A potential focus of classroom assessment at the outset of instruction is to elicit students' interests and experiences that may be relevant to the goals for instruction. However, identifying interests has not often been a focus of classroom assessment research in science, although it has been used to motivate and design assessments in specific curricula.[20]

One approach that could prove fruitful for classroom assessment is a strategy used in an elementary curriculum unit called *Micros and Me* (Tzou et al., 2007). The unit aims to engage students in the practice of argumentation to learn about key ideas in microbiology. In contrast to many curriculum units, however, this example provides students with the opportunity to pursue investigations related to issues that are relevant to them. The researchers adapted a qualitative

---

[19]The phrase "disposition to engage" is used in the context of science education to refer to students' degree of engagement with and motivation to persevere with scientific thinking.

[20]One example is *Issues, Evidence, and You*: see Science Education for Public Understanding Program (SEPUP) (1995) and Wilson and Sloane (2000).

methodology from psychology, photo-elicitation, which is used to identify these issues. Research participants take photos that become the basis for interviews that elicit aspects of participants' everyday lives (Clark-Ibañez, 2004). In *Micros and Me,* at the beginning of the unit, students take photos of things or activities they do to prevent disease and stay healthy. They share these photos in class, as a way to bring personally relevant experiences into the classroom to launch the unit. Their documentation also helps launch a student-led investigation focused on students' own questions, which are refined as students encounter key ideas in microbiology.

In describing the curriculum, Tzou and Bell (2010) do not call out the practice of self-documentation of students' personally relevant experiences as a form of assessment. At the same time, they note that a key function of self-documentation is to "elicit and make visible students' everyday expertise" relevant to the unit content (Tzou and Bell, 2010, p. 1136). Eliciting and making visible prior knowledge is an important aspect of assessment that is used to guide instruction. It holds promise as a way to identify diversity in the classroom in science that can be used to help students productively engage in science practices (Clark-Ibañez, 2004; Tzou and Bell, 2010; Tzou et al., 2007).

Professional Development

The framework emphasizes that professional development will be an indispensable component of the changes to science education it calls for (see National Research Council, 2012a, Ch. 10). The needed changes in instruction are beyond our charge, but in the context of classroom assessment, we note that significant adaptation will be asked of teachers. They will need systematic opportunities to learn how to use classroom discourse as a means to elicit, develop, and assess student thinking. The Contingent Pedagogies Project (see Example 4, above) illustrates one way to organize such professional development. In that approach, professional development included opportunities for teachers to learn how to orchestrate classroom discussion of core disciplinary ideas. Teachers also learned how to make use of specific discussion strategies to support the practice of argumentation.

Eliciting student thinking through skillful use of discussion is not enough, however. Tasks or teacher questions also have to successfully elicit and display students' problematic ways of reasoning about disciplinary core ideas and problematic aspects of their participation in practices. They must also elicit the interests and experiences students bring, so they can build on them throughout instruction. This is part of the process of integrating teaching and assessment. Thus, both teachers and assessment developers need to be aware of the typical student ideas about a topic and the various problematic alternative conceptions that students are likely to hold. (This is often called pedagogical content knowledge.) In addition, teachers need a system for interpreting students' responses to tasks or questions. That system should be intelligible and usable in practice: it cannot be so elaborate that teachers find it difficult to use in order to understand student thinking during instruction. (The construct map and its associated scoring guide shown in Chapter 3 are an example of such a system.)

CONCLUSIONS AND RECOMMENDATIONS

The primary conclusion we draw from these examples is that it is possible to design tasks and contexts in which teachers elicit students' thinking about disciplinary core ideas and crosscutting concepts by engaging them in scientific practices. Tasks designed with the characteristics we have discussed (three dimensions, interconnections among concepts and practices, a way to identify students' place on a continuum) produce artifacts, discussions, and activities that provide teachers with information about students' thinking and so can help them make decisions about how to proceed or how to adjust subsequent instruction or to evaluate the learning that took place over a specified period of time.

Questions have been raised about whether students can achieve the ambitious performance expectations in the NGSS. The implementation of the NGSS is a complex subject that is beyond the scope of our charge; however, each of the examples shown has been implemented with diverse samples of students,[21] and there have been students who succeeded on them (although there are also students who did not). The tasks in our examples assess learning that is part of a well-designed, coherent sequence of instruction on topics and in ways that are very similar to NGSS performance expectations. Each example offers multiple opportunities to engage in scientific practices and encourage students to draw connections among ideas, thus developing familiarity with crosscutting concepts.

CONCLUSION 4-1 **Tasks designed to assess the performance expectations in the Next Generation Science Standards will need to have the following characteristics:**

---

[21]Samples included students from rural and inner-city schools, from diverse racial and ethnic backgrounds, and English-language learners.

- multiple components that reflect the connected use of different scientific practices in the context of interconnected disciplinary ideas and crosscutting concepts;
- reflect the progressive nature of learning by providing information about where students fall on a continuum between expected beginning and ending points in a given unit or grade; and
- an interpretive system for evaluating a range of student products that is specific enough to be useful for helping teachers understand the range of student responses and that provides tools to helping them decide on next steps in instruction.

CONCLUSION 4-2 To develop the skills and dispositions to use scientific and engineering practices needed to further their learning and to solve problems, students need to experience instruction in which they (1) use multiple practices in developing a particular core idea and (2) apply each practice in the context of multiple core ideas. Effective use of the practices often requires that they be used in concert with one another, such as in supporting explanation with an argument or using mathematics to analyze data. Classroom assessments should include at least some tasks that reflect the connected use of multiple practices.

CONCLUSION 4-3 It is possible to design assessment tasks and scoring rubrics that assess three-dimensional science learning. Such assessments provide evidence that informs teachers and students of the strengths and weaknesses of a student's current understanding, which can guide further instruction and student learning and can also be used to evaluate students' learning.

We emphasize that implementing the conception of science learning envisioned in the framework and the NGSS will require teachers who are well trained in assessment strategies such as those discussed in this chapter. Professional development will be essential in meeting this goal.

CONCLUSION 4-4 Assessments of three-dimensional science learning are challenging to design, implement, and properly interpret. Teachers will need extensive professional development to successfully incorporate this type of assessment into their practice.

On the basis of the conclusions above, the committee offers recommendations about professional development and for curriculum and assessment development.

RECOMMENDATION 4-1 State and district leaders who design professional development for teachers should ensure that it addresses the changes called for by the framework and the Next Generation Science Standards in both the design and use of assessment tasks and instructional strategies. Professional development must support teachers in integrating practices, crosscutting concepts, and disciplinary core ideas in inclusive and engaging instruction and in using new modes of assessment that support such instructional activities.

Developing assessment tasks of this type will require the participation of several different kinds of experts. First, for the tasks to accurately reflect science ideas, scientists will need to be involved. Second, experts in science learning will also be needed to ensure that knowledge from research on learning is used as a guide to what is expected of students. Third, assessment experts will be needed to clarify relationships among tasks and the forms of knowledge and practice that the items are intended to elicit. Fourth, practitioners will be needed to ensure that the tasks and interpretive frameworks linked to them are usable in classrooms. And fifth, as we discuss further in Chapter 6, this multidisciplinary group of experts will need to include people who have knowledge of and experience with population subgroups, such as students with disabilities and students with varied cultural backgrounds, to ensure that the tasks are not biased for or against any subgroups of students for reasons irrelevant to what is being measured.

We note also that curricula, textbooks, and other resources, such as digital content, in which assessments may be embedded, will also need to reflect the characteristics we have discussed—and their development will present similar challenges. For teachers to incorporate tasks of this type into their practice, and to design additional tasks for their classrooms, they will need to have worked with many good examples in their curriculum materials and professional development opportunities.

RECOMMENDATION 4-2 Curriculum developers, assessment developers, and others who create resource materials aligned to the science framework and the Next Generation Science Standards should ensure that assessment activities included in such materials (such as mid- and end-of-chapter activities, suggested tasks for unit assessment, and online activities) require students to engage in practices that demonstrate their understanding of core ideas and crosscutting concepts. These materials should also reflect multiple dimensions of diversity (e.g., by connecting with students' cultural and linguistic identities). In designing these materials, development teams need to include experts in science, science learning, assessment design, equity and diversity, and science teaching.

5

ASSESSMENT FOR MONITORING

In Chapter 4, we focused on assessments that are used as a part of classroom instructional activities. In this chapter we turn to assessments that are distinct from classroom instruction and used to monitor or audit student learning over time. We refer to them as "monitoring assessments" or "external assessments."[1] They can be used to answer a range of important questions about student learning, such as: How much have the students in a certain school or school system learned over the course of a year? How does achievement in one school system compare with achievement in another? Is one instructional technique or curricular program more effective than another? What are the effects of a particular policy measure, such as reduction in class size? Table 5-1 shows examples of the variety of questions that monitoring assessments may be designed to answer at different levels of the education system.

The tasks used in assessments designed for monitoring purposes need to have the same basic characteristics as those for classroom assessments (discussed in Chapter 4) in order to align with the Next Generation Science Standards (NGSS): they will need to address the progressive nature of learning, include multiple components that reflect three-dimensional science learning, and include an interpretive system for the evaluation of a range of student products. In addition, assessments for monitoring need to be designed so that they can be given to large numbers of students, are sufficiently standardized to support the intended monitoring purpose (which may involve

---

[1]External assessments (sometimes referred to as large-scale assessments) are designed or selected outside of the classroom, such as by districts, states, countries, or international bodies, and are typically used to audit or monitor learning.

TABLE 5-1 Questions Answered by Monitoring Assessments

Types of inferences	Levels of the Education System			
	Individual Students	Schools or District	Policy Monitoring	Program Evaluation
Criterion-referenced	Have individual students demonstrated adequate performance in science?	Have schools demonstrated adequate performance in science this year?	How many students in state X have demonstrated proficiency in science?	Has program X increased the proportion of students who are proficient?
Longitudinal and comparative across time	Have individual students demonstrated growth across years in science?	Has the mean performance for the district grown across years? How does this year's performance compare to last year's?	How does this year's performance compare to last year's?	Have students in program X increased in proficiency across several years?
Comparative across groups	How does this student compare to others in the school/state?	How does school/district X compare to school/district Y?	How many students in different states have demonstrated proficiency in science?	Is program X more effective in certain subgroups?

high-stakes decisions about students, teachers, or schools), cover an appropriate breadth of the NGSS, and are cost-effective.

The measurement field has considerable experience in developing assessments that meet some of the monitoring functions shown in Table 5-1. In science, such assessments are typically composed predominantly of multiple-choice and short-answer, constructed-response questions. However, the sorts of items likely to be useful for adequately measuring the NGSS performance expectations—extended constructed-response questions and performance tasks—have historically posed challenges when used in assessment programs intended for system monitoring.

In this chapter we explore strategies for developing assessments of the NGSS that can be used for monitoring purposes. We begin with a brief look at currently used assessments, considering them in light of the NGSS. We next discuss the challenges of using performance tasks in assessments intended for administration on a large scale, such as a district, a state, or the national level, and we revisit the lessons learned from other attempts to do so. We then offer suggestions for approaches to using these types of tasks to provide monitoring data that are aligned with the NGSS, and we highlight examples of tasks and situations that can be used to provide appropriate forms of evidence, as well as some of the ways in which advances in measurement technology can support this work.

CURRENT SCIENCE MONITORING ASSESSMENTS

In the United States, the data currently used to answer monitoring-related questions about science learning are predominantly obtained through assessments that use two types of test administration (or data collection) strategies. The first is a *fixed-form test*, in which, on a given testing occasion, all students take the same form[2] of the test. The science assessments used by states to comply with the No Child Left Behind Act (NCLB) are examples of this test administration strategy: each public school student at the tested grade level in a given state takes the full test. According to NCLB requirements, these tests are given to all students in the state at least once in each of three grade spans (K-5, 6-8, 9-12). Fixed-form tests of all students (census tests) are designed to yield individual-level scores, which are used to address the questions about student-level performance shown in the first column of Table 5-1 (above). The scores are also aggregated as needed to provide information for the monitoring questions about school-, district-, and state-level performance shown in the three right-hand columns.

The second type of test administration strategy makes use of *matrix sampling*, which is used when the primary interest is group- or population-level estimates (i.e., schools or districts), rather than individual-level estimates. No individual student takes the full set of items and tasks. Instead, each of the tasks is completed by a sample of students that is sufficiently representative to yield valid and reliable scores for schools, states, or the nation. This method makes it possible to gather data on a larger and more representative collection of items or tasks for a given topic than any one student could be expected to complete in the time allocated for testing. In some applications, all students from a school or district are tested (with different parts of the whole test). In other applications, only some students are sampled for testing, but in sufficient number and representativeness that the results will provide an accurate estimate of how the entire school or district would perform.

---

[2]A *test form* is a set of assessment questions typically given to one or more students as part of an assessment administration.

The best-known example in the United States of an assessment that makes use of a matrix-sampling approach is the National Assessment of Educational Progress (NAEP), known as the "Nation's Report Card." NAEP is given to representative samples of 4th, 8th, and 12th graders, with the academic subjects and grade levels that are assessed varying from year to year.[3] The assessment uses matrix sampling of items to cover the full spectrum of each content framework (e.g., the NAEP science framework) in the allotted administration time. The matrix-sampling approach used by NAEP allows reporting of group-level scores (including demographic subgroups) for the nation, individual states, and a few large urban districts, but the design does not support reporting of individual-level or school-level scores. Thus, NAEP can provide data to answer some of the monitoring questions listed in Table 5-1, but not the questions in first or fourth columns. Matrix-sampling approaches have not generally been possible in the context of state testing in the last decade because of the requirements of NCLB for individual student reporting. When individual student results are not required, matrix sampling is a powerful and relatively straightforward option.

These two types of administration strategies for external assessments can be combined to answer different monitoring questions about student learning. When the questions require scores for individuals, generally all students are tested with fixed- or comparable test forms. But when group-level scores will suffice, a matrix-sampling approach can be used. Both approaches can be combined in a single test: for example, a test could include both a fixed-form component for estimating individual performance and a matrix-sampled component that is used to estimate a fuller range of performance at the school level. This design was used by several states prior to the implementation of NCLB, including Massachusetts, Maine, and Wyoming (see descriptions in National Research Council, 2010; Hamilton et al., 2002). That is, hybrid designs can be constructed to include a substantial enough fixed or common portion of the test to support individual estimates, with each student taking one of multiple matrix forms to ensure broad coverage at the school level.

The science tests that are currently used for monitoring purposes are not suitable to evaluate progress in meeting the performance expectations in the NGSS, for two reasons. First, the NGSS have only recently been published, so the current tests are not aligned with them in terms of content and the focus

---

[3]The schedule for NAEP test administrations is available at http://www.nagb.org/naep/assessment-schedule.html [November 2013].

on practices. Second, the current monitoring tests do not use the types of tasks that will be needed to assess three-dimensional science learning. As we discuss in Chapters 3 and 4, assessing three-dimensional science learning will require examining the way students perform scientific and engineering practices and apply crosscutting concepts while they are engaged with disciplinary core ideas.

Currently, some state science assessments include the types of questions that could be used for assessing three-dimensional learning (e.g., questions that make use of technology to present simulations or those that require extended constructed responses), but most rely predominantly on multiple-choice questions that are not designed to do so. In most cases, the items assess factual knowledge rather than application of core ideas or aspects of inquiry that are largely decoupled from core ideas. They do not use the types of multicomponent tasks that examine students' performance of scientific and engineering practices in the context of disciplinary core ideas and crosscutting concepts nor do they use tasks that reflect the connected use of different scientific practices in the context of interconnected disciplinary ideas and crosscutting concepts. Similarly, NAEP's science assessment uses some constructed-response questions, but these also are not designed to measure three-dimensional science learning. In 2009, NAEP administered a new type of science assessment that made use of interactive computer and hands-on tasks. These task formats are closer to what is required for measuring the NGSS performance expectations (see discussion below), but they are not yet aligned with the NGSS. Consequently, current external assessments cannot readily be used for monitoring students' progress in meeting the NGSS performance expectations.

We note, however, that NAEP is not a static assessment program. It periodically undertakes major revisions to the framework used to guide the processes of assessment design and task development. NAEP is also increasingly incorporating technology as a key aspect of task design and assessment of student performance. The next revision of the NAEP science framework may bring it into closer alignment with the framework and the NGSS. Thus, the NAEP science assessment might ultimately constitute an effective way to monitor the overall progress of science teaching and learning in America's classrooms in ways consistent with implementation of the framework and the NGSS.

INCLUDING PERFORMANCE TASKS IN MONITORING ASSESSMENTS

Implementation of the NGSS provides an opportunity to expand the ways in which science assessment is designed and implemented in the United States and the ways in which data are collected to address the monitoring questions shown in

Table 5-1. We see two primary challenges to taking advantage of this opportunity. One is to design assessment tasks so that they measure the NGSS performance expectations. The other is to determine strategies for assembling these tasks into assessments that can be administered in ways that produce scores that are valid, reliable, and fair and meet the particular technical measurement requirements necessary to support an intended monitoring purpose.

Measurement and Implementation Issues

In Chapter 3, we note that the selection and development of assessment tasks should be guided by the constructs to be assessed and the best ways of eliciting evidence about a student's proficiency relative to that construct. The NGSS performance expectations emphasize the importance of providing students the opportunity to demonstrate their proficiencies in both science content and practices. Ideally, evidence of those proficiencies would be based on observations of students actually engaging in scientific and engineering practices relative to disciplinary core ideas. In the measurement field, these types of assessment tasks are typically performance based and include questions that require students to construct or supply an answer, produce a product, or perform an activity. Most of the tasks we discuss in Chapters 2, 3, and 4 are examples of performance tasks.

Performance tasks can be and have been designed to work well in a classroom setting to help guide instructional decisions making. For several reasons, they have been less frequently used in the context of monitoring assessments administered on a large scale.

First, monitoring assessments are typically designed to cover a much broader domain than tests used in classroom settings. When the goal is to assess an entire year or more of student learning, it is difficult to obtain a broad enough sampling of an individual student's achievement using performance tasks. But with fewer tasks, there is less opportunity to fully represent the domain of interest.

Second, the reliability, or generalizability, of the resulting scores can be problematic. Generalizability refers to the extent to which a student's test scores reflect a stable or consistent construct rather than error and supports a valid inference about students' proficiency with respect to the domain being tested. Obtaining reliable individual scores requires that students each take multiple performance tasks, but administering enough tasks to obtain the desired reliability often creates feasibility problems in terms of the cost and time for testing. Careful task and test design (described below) can help address this issue.

Third, some of the monitoring purposes shown in Table 5-1 (in the second row) require comparisons across time. When the goal is to examine performance across time, the assessment conditions and tasks need to be comparable across the two testing occasions. If the goal is to compare the performance of this year's students with that of last year's students, the two groups of students should be required to respond to the same set of tasks or a different but equivalent set of tasks (equivalent in terms of difficulty and content coverage). This requirement presents a challenge for assessments using performance tasks since such tasks generally cannot be reused because they are based on situations that are often highly memorable.[4] And, once they are given, they are usually treated as publicly available.[5] Another option for comparison across time is to give a second group of students a different set of tasks and use statistical equating methods to adjust for differences in the difficulty of the tasks so that the scores can be placed on the same scale.[6] However, most equating designs rely on the reuse of some tasks or items. To date, the problem of equating assessments that rely *solely* on performance tasks has not yet been solved. Some assessment programs that include both performance tasks and other sorts of items use the items that are not performance based to equate different test forms, but this approach is not ideal—the two types of tasks may actually measure somewhat different constructs, so there is a need for studies that explore when such equating would likely yield accurate results.

Fourth, scoring performance tasks is a challenge. As we discuss in Chapter 3, performance tasks are typically scored using a rubric that lays out criteria for assigning scores. The rubric describes the features of students' responses required for each score and usually includes examples of student work at each scoring level. Most performance tasks are currently scored by humans who are trained to apply the criteria. Although computer-based scoring algorithms are increasingly in use, they are not generally used for content-based tasks (see, e.g., Bennett and Bejar, 1998; Braun et al., 2006; Nehm and Härtig, 2011; Williamson et al., 2006, 2012). When humans do the scoring, their variability in applying the criteria

---

[4]That is, test takers may talk about them after the test is completed, and share them with each other and their teachers. This exposes the questions and allows other students to practice for them or similar tasks, potentially in ways that affect the ability of the task to measure the intended construct.

[5]For similar reasons, it can be difficult to field test these kinds of items.

[6]For a full discussion of equating methods, which is beyond the scope of this report, see Kolen and Brennan (2004) or Holland and Dorans (2006).

introduces judgment uncertainty. Using multiple scorers for each response reduces this uncertainty, but it adds to the time and cost required for scoring.

This particular form of uncertainty does not affect multiple-choice items, but they are subject to uncertainty because of guessing, something that is much less likely to affect performance tasks. To deal with these issues, a combination of response types could be used, including some that require demonstrations, some that require short constructed responses, and some that use a selected-response format. Selected-response formats, particularly multiple-choice questions, have often been criticized as only being useful for assessing low-level knowledge and skills. But this criticism refers primarily to isolated multiple-choice questions that are poorly related to an overall assessment design. (Examples include questions that are not related to a well-developed construct map in the construct-modeling approach or not based on the claims and inferences in an evidence-centered design approach; see Chapter 3). With a small set of contextually linked items that are closely related to an assessment design, the difference between well-produced selected-response items and open-ended items may not be substantial. Using a combination of response types can help to minimize concerns associated with using only performance tasks on assessments intended for monitoring purposes.

Examples

Despite the various measurement and implementation challenges discussed above, a number of assessment programs have made use of performance tasks and port-folios[7] of student work. Some were quite successful and are ongoing, and some experienced difficulties that led to their discontinuation. In considering options for assessing the NGSS performance expectations for monitoring purposes, we began by reviewing assessment programs that have made use of performance tasks, as well as those that have used portfolios. At the state level, Kentucky, Vermont, and Maryland implemented such assessment programs in the late 1980s and early 1990s.

In 1990, Kentucky adopted an assessment for students in grades 4, 8, and 11 that included three types of questions: multiple-choice and short essay questions, performance tasks that required students to solve practical and applied

---

[7]A portfolio is a collection of work, often with personal commentary or self-analysis, that is assembled over time as a cumulative record of accomplishment (see Hamilton et al., 2009). A portfolio can be either standardized or nonstandardized: in a standardized portfolio, the materials are developed in response to specific guidelines; in a nonstandardized portfolio, the students and teachers are free to choose what to include.

problems, and portfolios in writing and mathematics in which students presented the best examples of their classroom work for a school year. Assessments were given in seven areas: reading, writing, social science, science, math, arts and humanities, and practical living/vocational studies. Scores were reported for individual students.

In 1988, Vermont implemented a statewide assessment in mathematics and writing for students in grades 4 and 8 that included two parts: a portfolio component and uniform subject-matter tests. For the portfolio, the tasks were not standardized: teachers and students were given unconstrained choice in selecting the products to be in them. The portfolios were complemented by subject-matter tests that were standardized and consisted of a variety of item types. Scores were reported for individual students.

The Maryland School Performance Assessment System (MSPAP) was implemented in 1991. It assessed reading, writing, language usage, mathematics, science, and social sciences in grades 3, 5, and 8. All of the tasks were performance based, including some that required short-answer responses and others that required complex, multistage responses to data, experiences, or text. Some of the activities integrated skills from several subject areas, some were hands-on tasks involving the use of equipment, and some were accompanied by preassessment activities that were not scored. The MSPAP used a matrix-sampling approach: that is, the items were sampled so that each student took only a portion of the exam in each subject. The sampling design allowed for the reporting of scores for schools but not for individual students.

These assessment programs were ambitious, innovative responses to calls for education reform. They made use of assessment approaches that were then cutting edge for the measurement field. They were discontinued for many reasons, including technical measurement problems, practical reasons (e.g., the costs of the assessments and the time they took to administer), as well as imposition of the accountability requirements of NCLB (see Chapter 1), which they could not readily satisfy.[8]

---

[8]A thorough analysis of the experiences in these states is beyond the scope of this report, but there have been several studies. For Kentucky, see Hambleton et al. (1995), Catterall et al. (1998). For Vermont, see Koretz et al. (1992a,b, 1993a,b, 1993c, 1994). For Maryland, see Hambleton et al. (2000), Ferrara (2009), and Yen and Ferrara (1997). Hamilton et al. (2009) provides an overview of all three of these programs. Hill and DePascale (2003) have pointed out that some critics of these programs failed to distinguish between the reliability of student-level scores and school-level scores. For purposes of school-level reporting, the technical quality of some of these assessments appears to have been better than generally assumed.

Other programs that use performance tasks are ongoing. At the state level, the science portion of the New England Common Assessment Program (NECAP) includes a performance component to assess inquiry skills, along with questions that rely on other formats. The state assessments in New York include laboratory tasks that students complete in the classroom and that are scored by teachers. NAEP routinely uses extended constructed-response questions, and in 2009 conducted a special science assessment that focused on hands-on tasks and computer simulations. The Program for International Student Assessment (PISA) includes constructed-response tasks that require analysis and applications of knowledge to novel problems or contexts. Portfolios are currently used as part of the advanced placement (AP) examination in studio art.

Beyond the K-12 level, the Collegiate Learning Assessment makes use of performance tasks and analytic writing tasks. For advanced teacher certification, the National Board for Professional Teaching Standards uses an assessment composed of two parts—a portfolio and a 1-day exam given at an assessment center.[9] The portfolio requires teachers to accumulate work samples over the course of a school year according to a specific set of instructions. The assessment center exam consists of constructed-response questions that measure the teacher's content and pedagogical knowledge. The portfolio and constructed responses are scored centrally by teachers who are specially trained.

The U.S. Medical Licensing Examination uses a performance-based assessment (called the Clinical Skills Assessment) as part of the series of exams required for medical licensure. The performance component is an assessment of clinical skills in which prospective physicians have to gather information from simulated patients, perform physical examinations, and communicate their findings to patients and colleagues.[10] Information from this assessment is considered along with scores from a traditional paper-and-pencil test of clinical skills in making licensing decisions.

Implications for Assessment of the NGSS

The experiences to date suggest strategies for addressing the technical challenges posed by the use of performance tasks in assessments designed for monitoring. In

---

[9]For details, see http://www.nbpts.org [June 2013].

[10]The assessment is done using "standardized patients" who are actors trained to serve as patients and to rate prospective physicians' clinical skills: for details, see http://www.usmle.org/step-2-cs/ [November 2013].

particular, much has been written about the procedures that lead to high-quality performance assessment and portfolios (see, e.g., Baker, 1994; Baxter and Glaser, 1998; Dietel, 1993; Dunbar et al., 1991; Hamilton et al., 2009; Koretz et al., 1994; Pecheone et al., 2010; Shavelson et al., 1993; Stiggins, 1987). This large body of work has produced important findings, particularly on scoring processes and score reliability.

With regard to the scoring process, particularly human scoring, strategies that can yield acceptable levels of interrater reliability include the following:

- use of standardized tasks that are designed with a clear idea of what constitutes poor and good performance;
- clear scoring rubrics that minimize the degree to which raters must make inferences as they apply the criteria to student work and that include several samples of student responses for each score level;
- involvement of raters who have significant knowledge of the skills being measured and the rating criteria being applied; and
- providing raters with thorough training, combined with procedures for monitoring their accuracy and guiding them in making corrections when inaccuracies are found.

With regard to score generalizability (i.e., the extent to which the score results for one set of tasks generalize to performance on another set of tasks), studies show that a moderate to large number of performance tasks are needed to produce scores that are sufficiently reliable to support high-stakes judgments about students (Shavelson et al., 1993; Dunbar et al., 1991; Linn et al., 1996).[11] Student performance can vary substantially among tasks because of unique features of the tasks and the interaction of those features with students' knowledge and experience. For example, in a study on the use of hands-on performance tasks in science with 5th- and 6th-grade students, Stecher and Klein (1997) found that three 45 to 50-minute class periods were needed to yield a score reliability of 0.80.[12] For the mathematics portfolio used in Vermont, Klein et al. (1995) estimated that as many as 25 pieces of student work would have been needed to produce a score reliable enough to support high-stakes decisions about individual stu-

---

[11]When test results are used to make important, high-stakes decisions about students, a reliability of 0.90 or greater is typically considered appropriate.

[12]The reader is referred to the actual article for details about the performance tasks used in this study.

dents. However, it should be noted that Vermont's portfolio system was designed to support school accountability determinations, and work by Hill and DePascale (2003) demonstrated that reliability levels that might cause concern at the individual level can still support school-level determinations. We note that this difficulty is not unique to assessments that rely on performance tasks: a test composed of only a small number of multiple-choice questions would also not produce high score reliability, nor would it be representative of a construct domain as defined by the NGSS. Research suggests that use of a well-designed set of tasks that make use of multiple-response formats could yield higher levels of score reliability than exclusive reliance on a small set of performance tasks (see, e.g., Wilson and Wang, 1995).

The measurement field has not yet fully solved the challenge of equating the scores from two or more assessments relying on performance tasks, but some strategies are available (see, e.g., Draney and Wilson, 2008). As noted above, some assessment programs like the College Board's advanced placement (AP) exams use a combination of item types, including some multiple-choice questions (that can generally be reused), which can be of assistance for equating, provided they are designed with reference to the same or similar performance expectations. Other assessment programs use a strategy of "pre-equating" by administering all of the tasks to randomly equivalent groups of students, possibly students in another state (for details, see Pecheone and Stahl, n.d., p. 23). Another strategy is to develop a large number of performance tasks and publicly release all of them and then to sample from them for each test administration. More recently, researchers have tried to develop task shells or templates to guide the development of tasks that are comparable but vary in particular details, so that the shells can be reused. This procedure has been suggested for the revised AP examination in biology where task models have been developed based on application of evidence-centered design principles (see Huff et al., 2012). As with the NGSS, this exam requires students to demonstrate their knowledge through applying a set of science practices.

DESIGN OPTIONS

There is no doubt that developing assessments that include performance tasks and that can be used to monitor students' performance with respect to the NGSS will be challenging, but prior research and development efforts, combined with lessons learned from prior and current operational programs, suggest some strategies for addressing the technical challenges. New methods will be needed, drawing on both existing and new approaches. Technology offers additional options, such as the use of simulations or external data sets and built-in data analysis tools, as well as flexible translation and accommodation tools. But technology also adds its own set of new equity challenges. In this section we propose design options and examples that we think are likely to prove fruitful, although some will need further development and research before they can be fully implemented and applied in any high-stakes environment. The approaches we suggest are based on several assumptions about adequate assessment of the NGSS for monitoring purposes.

Assumptions

It will not be possible to cover all of the performance expectations for a given grade (or grade band) during a typical single testing session of 60-90 minutes. To obtain a sufficient estimate of a single student's proficiency with the performance expectations, multiple testing sessions would be necessary. Even with multiple testing sessions, however, assessments designed for monitoring purposes alone cannot fully cover the NGSS performance expectations for a given grade within a reasonable testing time and cost. Moreover, some performance expectations will be difficult to assess using tasks not tied directly to a school's curriculum and that can be completed in 90 minutes or less. Thus, our first assumption is that such assessments will need to include a combination of tasks given at a time mandated by the state or district (*on-demand assessment components*) and tasks given at a time that fits the instructional sequence in the classroom (*classroom-embedded assessment components*).

Second, we assume that assessments used for monitoring purposes, like assessments used for instructional support in classrooms, will include multiple types of tasks. That is, we assume that the individual tasks that compose a monitoring assessment will include varied formats: some that require actual demonstrations of practices, some that make use of short- and extended-constructed responses, and some that use carefully designed selected-response questions. Use of multiple components will help to cover the performance expectations more completely than any assessment that uses only one format.

We recognize that the approaches we suggest for gathering assessment information may not yield the level of comparability of results that educators, policy makers, researchers, and other users of assessment data have been accustomed to, particularly at the individual student level. Thus, our third assumption is that developing assessments that validly measure the NGSS is more important than achieving strict comparability. There are tradeoffs to be considered. Traditional approaches that have been shown to produce comparable results, which heavily rely on selected-response items, will not likely be adequate for assessing the full breadth and depth of the NGSS performance expectations, particularly in assessing students' proficiency with the application of the scientific and engineering practices in the context of disciplinary core ideas. The new approaches that we propose for consideration (see below) involve hybrid designs employing performance tasks that may not yield strictly comparable results, which will make it difficult to make some of the comparisons required for certain monitoring purposes.[13] We assume that users will need to accept different conceptualizations and degrees of comparability in order to properly assess the NGSS.[14]

Fourth, we assume that the use of technology can address some of the challenges discussed above (and below). For example, technology can be useful in scoring multiple aspects of students' responses on performance tasks, and technology-enhanced questions (e.g., those using simulations or data display tools) can be useful if not essential in designing more efficient ways for students to demonstrate their proficiency in engaging in some of the science practices. Nevertheless, technology alone is unlikely to solve problems of score reliability or of equating, among other challenges.

Finally, we assume that matrix sampling will be an important tool in the design of assessments for monitoring purposes to ensure that there is proper coverage of the broad domain of the NGSS. Matrix sampling as a design principle may be extremely important even when individual scores are needed as part of the monitoring process. This assumption includes hybrid designs in which all students respond to the same core set of tasks that are mixed with matrix-sampled tasks to ensure representativeness of the NGSS for monitoring inferences about student learning at higher levels of aggregation (see the second, third, and fourth columns in Table 5-1, above).

---

[13] A useful discussion of issues related to comparability can be found in Gong and DePascale (2013).

[14] We note that in the United States comparability is frequently based on a statistical (psychometric) concept; in other countries, comparability relies on a balance between psychometric evidence and evidence derived from assessment design information and professional judgment (i.e., expert judgment as to commonality across assessments in terms of the breadth, depth, and format of coverage). Examples include the United Kingdom system of assessment at the high school level and functions served by their monitoring body, called the Office of Qualifications and Examinations Regulations (Ofqual), to ensure comparability across different examination programs all tied to the same curricular frameworks. See http://ofqual.gov.uk/how-we-regulate/ [November 2013].

Two Classes of Design Options

With these assumptions in mind, we suggest two broad classes of design options. The first involves the use of on-demand assessment components and the second makes use of classroom-embedded assessment components. For each class, we provide a general description of options, illustrating the options with one or more operational assessment programs. For selective cases, we also provide examples of the types of performance tasks that might be used as part of the design option. It should be noted that our two general classes of design options are not being presented as an either-or contrast. Rather, they should be seen as options that might be creatively and selectively combined, with varying weighting, to produce a monitoring assessment that appropriately and adequately reflects the depth and breadth of the NGSS.

On-Demand Assessment Components

As noted above, one component of a monitoring system could include an on-demand assessment that might be administered in one or more sessions toward the end of a given academic year. Such an assessment would be designed to cover multiple aspects of the NGSS and might typically be composed of mixed-item formats with either written constructed responses or performance tasks or both.

Mixed-Item Formats with Written Responses

A mixed-item format containing multiple-choice and short and extended constructed-response questions characterizes certain monitoring assessments. As an example, we can consider the revised AP assessment for biology (College Board, 2011; Huff et al., 2010; Wood, 2009). Though administered on a large scale, the tests for AP courses are aligned to a centrally developed curriculum, the AP framework, which is also used to develop instructional materials for the course (College Board, 2011). Most AP courses are for 1 year, and students take a 3-hour exam at the end of the course. (Students are also allowed to take the exam without having taken the associated course.) Scores on the exam can be used to obtain college credit, as well as to meet high school graduation requirements.

Using the complementary processes of "backwards design" (Wiggins and McTighe, 2005) and evidence-centered design (see Chapter 3), a curriculum framework was developed for biology organized in terms of disciplinary big ideas, enduring understandings, and supporting knowledge, as well as a set of seven science practices. This structure parallels that of the core ideas and science practices in the K-12 framework. The AP biology curriculum framework focuses on the integration, or in the College Board's terminology, "fusion," of core scientific ideas with scientific practice in much the same way as the NGSS performance expectations. And like what is advocated in the K-12 science framework (see National Research Council, 2012a) and realized in the NGSS, a set of performance expectations or learning objectives was defined for the biology discipline. Learning objectives articulate what students should know and be able to do and they are stated in the form of claims, such as "the student is able to construct explanations of the mechanisms and structural features of cells that allow organisms to capture, store or use free energy" (learning objective 2.5). Each learning objective is designed to help teachers integrate science practices with specific content and to provide them with information about how students will be expected to demonstrate their knowledge and abilities (College Board, 2013a, p. 7). Learning objectives guide instruction and also serve as a guide for developing the assessment questions since they constitute the claim components in the College Board system for AP assessment development. Through the use of evidence-centered design, sets of claim-evidence pairs were elaborated in biology that guide development of assessment tasks for the new AP biology exam.

Assessment Task Example 9, Photosynthesis and Plant Evolution: An example task from the new AP biology assessment demonstrates the use of a mixed-item formats with written responses. As shown in Figure 5-1, this task makes use of both multiple-choice questions and free-response questions. The latter include both short-answer and extended constructed responses. It was given as part of a set of eight free-response questions (six short-answer questions and two extended constructed-response questions) during a testing session that lasted 90 minutes. The instructions to students suggested that this question would require 22 minutes to answer.

The example task has multiple components in which students make use of data in two graphs and a table to respond to questions about light absorption. It asks students to work with scientific theory and evidence to explain how the processes of natural selection and evolution could have resulted in different photosynthetic organisms that absorb light in different ranges of the visible light spectrum. Students were asked to use experimental data (absorption spectra) to identify two different photosynthetic pigments and to explain how the data support their identification. Students were then presented with a description of an experiment for investigating how the wavelength of available light affects the rate of photosynthe-

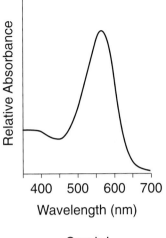

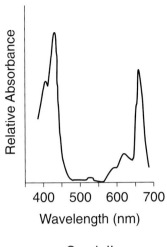

Graph I   Graph II

Color	Wavelength (nm)
Violet	380–450
Blue	450–475
Cyan	475–495
Green	495–570
Yellow	570–590
Orange	590–620
Red	620–750

FIGURE 5-1 AP biology example.
NOTE: See text for discussion.
SOURCE: College Board (2013a, p. 4).
Reprinted with permission.

sis in autotrophic organisms. Students were asked to predict the relative rates of photosynthesis in three treatment groups, each exposed to a different wavelength of light, and to justify their prediction using their knowledge and understanding about the transfer of energy in photosynthesis. Finally, students were asked to propose a possible evolutionary history of plants by connecting differences in resource availability with different selective pressures that drive the process of evolution through natural selection.

Collectively, the multiple components in this task are designed to provide evidence relevant to the nine learning objectives, which are shown in Box 5-1. The

BOX 5-1

LEARNING OBJECTIVES (LO) FOR SAMPLE AP BIOLOGY QUESTION

LO 1.12

The student is able to connect scientific evidence from many scientific disciplines to support the modern concept of evolution.

LO 1.13

The student is able to construct and/or justify mathematical models, diagrams or simulations that represent processes of biological evolution.

LO 1.2

The student is able to analyze data related to questions of speciation and extinction throughout the Earth's history.

LO 1.25

The student is able to describe a model that represents evolution within a population.

LO 2.24

The student is able to analyze data to identify possible patterns and relationships between a biotic or abiotic factor and a biological system (cells, organisms, populations, communities, or ecosystems).

LO 2.5

The student is able to construct explanations of the mechanisms and structural features of cells that allow organisms to capture, store or use free energy.

LO 4.4

The student is able to make a prediction about the interactions of subcellular organelles.

LO 4.5

The student is able to construct explanations based on scientific evidence as to how interactions of subcellular structures provide essential functions.

LO 4.6

The student is able to use representations and models to analyze situations qualitatively to describe how interactions of subcellular structures, which possess specialized functions, provide essential functions.

SOURCE: College Board (2011). Reprinted with permission.

task has a total point value of 10 and each component of the task (a, b, c) has an associated scoring rubric (see Figure 5-2). Note that in the case of responses that require an explanation or justification, the scoring rubric includes examples of the acceptable evidence in the written responses. Figure 5-3 shows two different student responses to this task: one in which the student earned all 10 possible points and one in which the student earned 6 points (3 points for Part a; 3 points for Part b; and 0 points for Part c).[15]

Mixed-Item Formats with Performance Tasks

Two current assessment programs use a mixed-item format with performance tasks. Both assessments are designed to measure inquiry skills as envisioned in the science standards that predate the new science framework and the NGSS. Thus, they are not fully aligned with the NGSS performance expectations. We highlight these two assessments not because of the specific kinds of questions that they use, but because the assessments require that students demonstrate science practices and interpret the results.

One assessment is the science component of the NECAP, used by New Hampshire, Rhode Island, and Vermont, and given to students in grades 4 and 8. The assessment includes three types of items: multiple-choice questions, short constructed-response questions, and performance tasks. The performance-based tasks present students with a research question. Students work in groups to conduct an investigation in order to gather the data they need to address the research question and then work individually to prepare their own written responses to the assessment questions.[16]

A second example is the statewide science assessment administered to the 4th and 8th grades in New York. The assessment includes both multiple-choice and performance tasks. For the performance part of the assessment, the classroom teacher sets up stations in the classroom according to specific instructions in the assessment manual. Students rotate from station to station to perform the task, record data from the experiment or demonstration, and answer specific

---

[15]Additional examples of student responses to this task, as well as examples of the other tasks, their scoring rubrics, and sample student responses, on the constructed-response section of the May 2013 exam can be found at http://apcentral.collegeboard.com/apc/members/exam/exam_information/1996.html [November 2013].

[16]Examples of questions are available at http://www.ride.ri.gov/InstructionAssessment/Assessment/NECAPAssessment/NECAPReleasedItems/tabid/426/LiveAccId/15470/Default.aspx [August 2013].

(a) **Identify** the pigment (chlorophyll *a* or bacteriorhodopsin) used to generate the absorption spectrum in each of the graphs above. **Explain** and **justify** your answer. (**3 points maximum**)

1 point per box
Identify BOTH pigments: Graph 1 = bacteriorhodopsin AND graph 2 = chlorophyll *a*
Explain that an organism containing bacteriorhodopsin appears purple because the pigment absorbs light in the green range of the light spectrum and/or reflects violet or red and blue light. The reflected red and blue light appears purple.
Explain that an organism containing chlorophyll *a* appears green because the pigment absorbs light in the red and blue ranges of the light spectrum and/or reflects green light.

(b) In an experiment, identical organisms containing the pigment from Graph II as the predominant light-capturing pigment are separated into three groups. The organisms in each group are illuminated with light of a single wavelength (650 nm for the first group, 550 nm for the second group, and 430 nm for the third group). The three light sources are of equal intensity, and all organisms are illuminated for equal lengths of time. **Predict** the relative rate of photosynthesis in each of the three groups. **Justify** your predictions. (**5 points maximum**)

Wavelength (Group)	Prediction (**1 point each box**)	Justification (**1 point each box**)
650 nm (1ˢᵗ Group)	Intermediate rate	An intermediate level of absorption occurs at 650 nm (compared to 430 nm and 550 nm); *therefore*, an intermediate amount of energy is available to drive photosynthesis.
550 nm (2ⁿᵈ Group)	Lowest rate	The lowest level of absorption occurs at 550 nm; *therefore*, the least amount of energy is available to drive photosynthesis.
430 nm (3ʳᵈ Group)	Highest rate	The highest level of absorption occurs at 430 nm; *therefore*, the greatest amount of energy is available to drive photosynthesis.

NOTE: A student who combines two groups (e.g., "the 650 nm and 430 nm groups have higher rates of photosynthesis compared to the 550 nm group") can earn a maximum of 4 points: up to 2 points for the prediction and up to 2 points for the justification.

(c) Bacteriorhodopsin has been found in aquatic organisms whose ancestors existed before the ancestors of plants evolved in the same environment. **Propose** a possible evolutionary history of plants that could have resulted in a predominant photosynthetic system that uses only some of the colors of the visible light spectrum. (**1 point per box; 2 points maximum**)

Proposal that includes an environmental selective pressure: <ul><li>Green light was being absorbed by aquatic organisms using bacteriorhodopsin.</li><li>Unabsorbed wavelengths of light were available resources that organisms could exploit.</li><li>Absorbing visible light at all wavelengths may provide too much energy to the organism.</li><li>Absorbing light from ultraviolet wavelengths (shorter wavelengths = higher energy) could cause damage to the organism.</li><li>Absorbing light with longer wavelengths may not provide sufficient energy for the organism.</li></ul>
Appropriate reasoning to support the proposal: <ul><li>Natural selection favored organisms that rely on pigments that absorb available wavelengths of light.</li><li>Endosymbiosis: chloroplasts evolved from cyanobacteria with pigments that used only certain wavelengths.</li><li>Genetic drift eliminated pigments that absorbed certain wavelengths of light.</li><li>Mutation(s) altered the pigment(s) used by organism.</li></ul>

FIGURE 5-2 Scoring rubrics for each part of AP biology example.
SOURCE: College Board (2013b, pp. 1-2). Reprinted with permission.

ANSWER PAGE FOR QUESTION 2

(a) In graph I bacteriorhodoprin is used to generate the absorbtion spectrum. This is known because Graph I shows a relatively low absorbtion rate for the color violet which is wavelength 380-450. Bacteriorhodoprin is usually found in purple photosynthetic bacteria and since the organism is purple, then it reflects rather than absorbs purple light. Graph II shows the absorbtion spectrum for chlorophyll a because it shows a low level of absorbtion of wavelengths 490-650 which correlates with the wavelength of green light (495-570). Chlorophyll a is found in green plants which means that green light would not be absorbed it would be reflected. Therefore, Graph II would represent chlorophyll a due to its low absorbtion of green light.

(b) The second group of organisms illuminated by 550nm light will have the lowest level of photosynthesis. This is because the main light-capturing pigment has a low absorbance of light in 550nm. Photosynthesis will be slow because the photosystems will not be able to capture enough light to excite the electrons and produce ATP and NADPH, the products of the light dependent reaction. The first group of organisms illuminated by 650nm will have a higher rate of photosynthesis than the second group

FIGURE 5-3 Two sample AP biology responses.
SOURCE: College Board (2013b, pp. 3-5, 8-9). Reprinted with permission.

$\angle A_2$

but lower than the third group. The absorption spectrum of the predominant light-capturing pigment absorbs more light at 650 nm than at 550 nm. The organisms will be able to absorb more light than the second group and be able to send more NADPH and ATP from the light-dependant reactions to the light-independant reactions, also known as the Calvin Cycle. The third group that is illuminated by 430 nm will have the highest rate of photosynthesis because 430 nm light is absorbed relatively easier than the other two wavelengths of light. The organisms in the third group will be able to absorb more light and therefore create more NADPH and ATP which will then cause more products of the Calvin cycle to form. The third group of organisms will also produce the most amount of oxygen.

(e) In an aquatic environment a plant would have access to mostly blue and cyan colors of light of around 450-495 nm wavelength. If the plant contained many pigments that absorbed red light, that ~~it barely~~ a plant could rarely gain access to,

FIGURE 5-3 Continued

Developing Assessments for the Next Generation Science Standards

ADDITIONAL PAGE FOR ANSWERING QUESTION 2

the plant would not be able to absorb enough light to undergo photosynthesis. The plant would most likely not be able to reproduce before it died. But if a plant had many photo pigments that could absorb blue light (which is plentiful in its environment) then the plant would thrive. It could pass on its genes and its offspring would have a higher fitness than plants that could only absorb red light. The remaining plants would only use the blue colors of the visible light spectrum because it wouldn't be efficient to have photo pigments that could absorb red lights.

FIGURE 5-3 Continued

ANSWER PAGE FOR QUESTION 2

a. The bacterorhadopsin ^(pigment) was used to generate Graph I because
(b) on the interval (350 nm, 600 nm) it has a relative minimum
for absorption near 450 nm which is included in the wavelength
range for violet. This means that at this wavelength the light
was not absorbed but reflected, giving the appearance of the
color at that wavelength. This pigment is found in purple
photosynthetic bacteria. The chlorophyll a was used to generate
Graph II, as chlorophyll is found in green plants, meaning that
green is reflected and not absorbed. In this graph, relative
absorbance is near 0 at 550 nm, which is included in the
wavelength range for green (495 nm - 570 nm).

b. The 3rd group will have the fastest rate of photosynthesis,
as the graph shows a peak of absorbance at approximately
450 nm. This is at the violet end of the light spectrum and is
the furthest in the graph from the green wavelength. The second
fastest rate of photosynthesis would occur in the first group, as
is represented on the graph and in the fact that this
wavelength is on the red end of the spectrum and
the second furthest from green in the graph. The second
group would have a very slow rate of photosynthesis. The
graph shows an absorbance of nearly zero as this wavelength
is in the green area of the spectrum, which is not absorbed by

FIGURE 5-3 Continued

questions.[17] In addition to these state programs, it is worth considering an international example of how a performance task can be included in a monitoring assessment.

Assessment Task Example 10, Floating and Sinking: To create standards for science education in Switzerland, a framework was designed that is similar to the U.S. framework. Assessments aligned with the Swiss framework were developed and administered to samples of students in order to obtain empirical data for specifying the standards. Like the U.S. framework, the Swiss framework defined three dimensions of science education—which they called skills, domains, and levels—and emphasized the idea of three-dimensional science learning. The domain dimension includes eight themes central to science, technology, society, and the environment (e.g., motion, force and energy, structures and changes of matter, ecosystems). The skills dimension covers scientific skills similar to the scientific practices listed in the U.S. framework. For each skill, several subskills are specified. For the skill "to ask questions and to investigate," five subskills are defined: (1) to look at phenomena more attentively, to explore more precisely, to observe, to describe, and to compare; (2) to raise questions, problems, and hypothesis; (3) to choose and apply suitable tools, instruments, and materials; (4) to conduct investigations, analyses, and experiments; and (5) to reflect on results and examination methods (see Labudde et al., 2012).

To collect evidence about student competence with respect to the framework, Swiss officials identified a set of experts to develop assessments. From the outset, this group emphasized that traditional approaches to assessment (e.g., paper-and-pencil questions assessing factual knowledge and simplistic understandings) would not be sufficient for evaluating the integrated learning reflected by the combinations of domains and skills in the framework. As a result, the group decided to follow the example of the Trends in Mathematics and Science Study (TIMSS), which (in its 1995 iteration) included an add-on study that used performance tasks to assess students' inquiry skills in 21 countries (Harmon et al., 1997). One of the performance tasks used for defining standards in science education in Switzerland is shown in Figure 5-4, and, for use in this report, has been translated from German to English.

---

[17]Examples are available at http://www.nysedregents.org/Grade4/Science/home.html and http://www.nysedregents.org/Grade8/Science/home.html [August 2013].

Floating and Sinking

You have

| One ship | Two large discs (each weighing 10 grams) | Two small discs (each weighing 4 grams) | A candle |

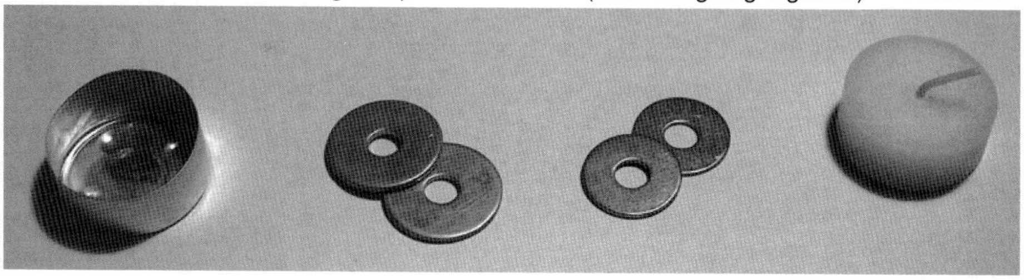

Your ship can be loaded in different ways. We will try out one way.

Question 1

One small disc is placed as cargo in the ship. The disc is placed on the inside edge of the ship, not in the center. What will happen when you put the ship in the water?

In the space below, draw a picture of what you think will happen. On the lines below, write an explanation of what you think will happen.

Scoring Rubric for Question 1

3 Points:
Drawing/answer that reflects the following ideas: The ship is floating but is tilted to one side. The placement of the disc on the inside edge of the ship caused the ship to float unevenly.

FIGURE 5-4 Sample performance-based task.

NOTES: The English translation of the three examples of answers are as follows "the little boat is heavy on the one side (at code 2)"; "it remains on the top of the water, but the little boat is inclined and water is coming in" (code 1, drawing on the left); "it tilts over" (code 1, drawing on the right).

SOURCE: Labudde et al. (2012). Copyright by the author; used with permission.

2 Points:
Drawing/answer that reflects the following concept: The ship is floating but is tilted to one side. There is no explanation for why it tilts.

1 Point:
Drawing/answer that indicates that the ship floats, but there is no recognition that the off-center placement of the weight causes the ship to float unevenly.

0 Points:
Drawing/answer that indicates that the ship sinks—or other answers/drawings.

Question 2

a. Place the disc in the ship as was demonstrated for question 1.
b. Place the ship onto the water.
c. Observe what happens.
d. In the space below, draw a picture of what happened. On the lines below, write an explanation of what happened.

Scoring Rubric for Question 2

2 Points:
The drawing contains the following elements: the water surface, the ship floating tilted in the water, the lowest point of the ship is the side containing the disc. The written explanation indicates that the ship floats but is tilted.

1 Point:
The drawing contains some points of the correct solution (e.g., it may contain two elements, such as the water surface and tilted ship, but part of the explantion is missing).

0 Points:
Other

FIGURE 5-4 Continued

Example   Responses

2 Points	*Das kleine ding macht auf der eien seite schwer*	Translation: The disc makes the ship heavy on one side.
1 Point	*es bleibt oben aber Das schifchen ist schreg und es kommt wasser rein.*	Translation: The ship floats but tilts and water comes in.
1 Point	*es kibt*	Translation: It turns over.
0 Points	*Es get immer an den rand.*	Translation: It constantly moves to the edge.

FIGURE 5-4 Continued

Question 3

What else would you like to know about the ship and what happens when it is loaded with the discs? Write your question below.

_____

_____

_____

Scoring Rubric for Question 3

Types of Questions:
a. Does the ship sink when I load it evenly with all four discs?
b. What happens if I load the ship with two large discs?

3 Points:
Question or hypotheses of type a

2 Points:
Question or hypotheses of type b

1 Points:
No real question/question not related to material/problem recognizable

0 Points:
Other questions (e.g., How far does it splash when I throw the discs into the water?) or statements (e.g., Put the disc into) the ship.

Question 4

Research your question. Perform an experiment to find the answer to your question. Draw and write down what you have found out.

FIGURE 5-4 Continued

Scoring Rubric for Question 4

2 Points

Answer fullfills the following criteria:

a. Tight relation to question: Design provides answer to the the posed question/problem.
b. The observations (drawing and text together) are detailed (e.g., The ship tilted to the left, the load fell off and sank quickly).

1 Point

Answer fullfills the following criteria:

a. Somewhat connected to the question: Design is at least directed toward the posed question/problem.
b. The observations (drawing and text together) are understandable but incomplete or not detailed (e.g., The ship tilted).

0 Points

Other answers

Question 5

Consider what you could learn from the experiments you have just done. Mark "Learned" if the statement indicates something you could find out from these experiments. Mark "Not Learned" if it is something you could not learn from these experiments.
[Below, the correct answers are indicated with an X.]

Learned	Not Learned	
X		When discs are placed at the edge of a ship, it can turn over and sink
	X	Ships need a motor.
X		The heavier a ship is, the deeper it sinks into the water.
X		A ship made from metal can be loaded with iron and still float.
	X	Round ships float better than long ships.

FIGURE 5-4 Continued

This task was one of several designed for use with students in 2nd grade. As part of the data collection activities, the tasks were given to 593 students; each student responded to two tasks and were given 30 minutes per task. The task was designed primarily to assess the student's skills in asking questions and investigating (more specifically, to look at phenomena more attentively, to explore more precisely, to observe, to describe, and to compare), within the domain of "motion, force and energy": for this task, the focus was on floating and sinking, or buoyancy in different contexts. The full task consisted of eight questions. Some of the questions involved placing grapes in water; other questions involved loading weights in a small "ship" and placing it in water. Figure 5-4 shows an excerpt from the portion of the task that involves the ship (see Table 1-1 for the specific disciplinary core ideas, scientific practices, and crosscutting concepts assessed).

In the excerpt shown in the figure, the first two activities ask students to observe a weighted ship floating on water and to describe their observations. Students were given a cup half full of water, a small ship, four metal discs (two large discs and two small discs), and a candle.[18] Students were instructed to (1) place the metal discs in the ship so they rested on the inside edge of the ship (i.e., off center); (2) place the ship into the water; (3) observe what happens; and (4) draw and describe in writing what they observed. The test proctor read the instructions out loud to the students and demonstrated how the discs should be placed in the ship and how the ship should be put into the water. The task included two additional activities. For one, students were asked to formulate a question and carry out an experiment to answer it. In the final section of the task, students were asked a series of questions about the type of information that could be learned from the experiments in the tasks. The figure shows the rubric and scoring criteria for the open-ended questions and the answer key for the final question. Sample responses are shown for the second activity.

As can be seen from this example, the task is about buoyancy, but it does not focus on assessing students' knowledge about what objects float (or do not float) or why objects float (or do not float). It also does not focus on students' general skill in observing a phenomenon and describing everything that has been observed. Instead, students had to recognize that the phenomenon to observe is about floating and sinking—more specifically, that when weights are placed off center in the ship, they cause the ship to float at an inclined angle or even to sink. Moreover, they were expected to recognize the way in which the off-center

---

[18]The candle could be used for other questions in the task.

load will cause the ship to tilt in the water. The task was specifically focused on the integration of students' knowledge about floating and sinking with their skill in observing and describing the key information. And the scoring criteria were directed at assessing students' ability to observe a phenomenon based on what they know about the phenomenon (i.e., what characteristics are important and how these characteristics are related to each other). Thus, the task provides an example of a set of questions that emphasize the integration of core ideas, crosscutting concepts, and practices.

Design of Performance Events

Drawing from the two state assessment program examples and the international assessment task example, we envision that this type of assessment, which we refer to as a "performance event," would be composed of a set of tasks that center on a major science question. The task set could include assessment questions that use a variety of formats, such as some selected-response or short-answer questions and some constructed-response questions, all of which lead to producing an extended response for a complex performance task. The short-answer questions would help students work through the steps involved in completing the task set. (See below for a discussion of ways to use technological approaches to design, administer, and score performance events.)

Each of the performance events could be designed to yield outcome scores based on the different formats: a performance task, short constructed-response tasks, and short-answer and selected-response questions. Each of these would be related to one or two practices, core ideas, or crosscutting concepts. A performance event would be administered over 2 to 3 days of class time. The first day could be spent on setting up the problem and answering most or all of the short- and long-answer constructed-response questions. This session could be timed (or untimed). The subsequent day(s) would be spent conducting the laboratory (or other investigation) and writing up the results.

Ideally, three or four of these performance assessments would be administered during an academic year, which would allow the task sets to cover a wide range of topics. The use of multiple items and multiple response types would help to address the reliability concerns that are often associated with the scores reported for performance tasks (see Dunbar et al., 1991). To manage implementation, such assessments could be administered during different "testing windows" during the spring or throughout the school year.

Use of multiple task sets also opens up other design possibilities, such as using a hybrid task sampling design (discussed above) in which all students at a grade level receive one common performance task, and the other tasks are given to different groups of students using matrix sampling. This design allows the common performance task to be used as a link for the matrix tasks so that student scores could be based on all of the tasks they complete. This design has the shortcoming of focusing the link among all the tasks on one particular task—thus opening up the linkage quality to weaknesses due to the specifics of that task. A better design would be to use all the tasks as linking tasks, varying the common task across many classrooms. Although there are many advantages to matrix-sampling approaches, identifying the appropriate matrix design will take careful consideration. For example, unless all the performance tasks are computer-based, the logistical and student-time burden of administering multiple tasks in the same classroom could be prohibitive. There are also risks associated with using all the tasks in an assessment in each classroom, such as security and memorability, which could limit the reuse of the tasks for subsequent assessments.[19]

The assessment strategies discussed above have varying degrees of overlap with the assessment plans that are currently in place for mathematics and language arts in the two Race to the Top Assessment Program consortia, the Partnership for Assessment of Readiness for College and Careers and the Smarter Balanced Assessment Consortium (see Chapter 1). Both are planning to use a mixed model with both performance tasks and computer-based selected-response and construct-response tasks (K-12 Center at Educational Testing Service, 2013). The different task types will be separated in time with respect to administration and in most grades the total testing time will be 2 or more hours.

Classroom-Embedded Assessment Components

As noted above, one component of a monitoring system could involve classroom-embedded tasks and performances that might be administered at different times in a given academic year so as to align with the completion of major units of instruction. These instructional units and assessments would be targeted at various sets of standards, such as those associated with one or more core ideas in the life sciences. Such a classroom-embedded assessment would be designed to cover more selective aspects of the NGSS and would be composed of tasks that require written constructed responses, performance activities, or both. We discuss three

---

[19]This format can also be viewed in terms of "replacement units": see discussion below.

options that involve the use of classroom-embedded assessment activities: replacement units, collections of performance tasks, and portfolios of work samples and projects

Replacement Units

Replacement units are curricular units that have been approved centrally (by the state or district) and made available to schools. They cover material or concepts that are already part of the curriculum, but they teach the material in a way that addresses the NGSS and promotes deeper learning. They are not intended to add topics to the existing curriculum, but rather to replace existing units in a way that is educative for teachers and students. The idea of replacement units builds from Marion and Shepard (2010).

Given the huge curricular, instructional, and assessment challenges associated with implementing the NGSS, replacement units would be designed to be used locally as meaningful examples to support capacity to implement the NGSS, as well as to provide evidence of student performance on the NGSS. The end-of-unit standardized assessment in the replacement unit would include performance tasks and perhaps short constructed-response tasks that could be used to provide data for monitoring student performance. The assessments could be scored locally by teachers or a central or regional scoring mechanism could be devised.

The units could be designed, for instance, by state consortia, regional labs, commercial vendors, or other groups of educators and subject-matter experts around a high-priority topic for a given grade level. Each replacement unit would include instructional supports for educators, formative assessment probes, and end-of-unit assessments. The supports embedded in the replacement units would serve as a useful model for trying to improve classroom assessment practices at a relatively large scale. In addition, the end-of-unit assessments, although not necessarily useful for short-term formative purposes, may serve additional instructional uses that affect the learning of future students or even for planning changes to instruction or curriculum for current students after the unit has been completed.

Collections of Performance Tasks

A second option would be for a state or district (or its contractors) to design standardized performance tasks that would be made available for teachers to use as designated points in curriculum programs. Classroom teachers could be trained to score these tasks, or student products could be submitted to the district or state and scored centrally. Results would be aggregated at the school, district, or state level to support monitoring purposes.

This option builds on an approach that was until recently used in Queensland, Australia, called the Queensland Comparable Assessment Tasks (QCATs). The QCAT consists of performance tasks in English, mathematics, and science that are administered in grades 4, 6, and 9. They are designed to engage students in solving meaningful problems. The structure of the Queensland system gives schools and teachers more control over assessment decisions than is currently the case in the United States. Schools have the option of using either centrally devised QCATs, which have been developed by the Queensland Studies Authority (QSA), with common requirements and parameters and graded according to a common guide, or school-devised tasks, which are developed by schools in accord with QSA design specifications.

The QCATs are not on-demand tests (i.e., not given at a time determined by the state); schools are given a period of 3-4 months to administer, score, and submit the scores to the QSA. The scores are used for low-stakes purposes.[20] Individual student scores are provided to teachers, students, and parents for instructional improvement purposes. Aggregate school-level scores are reported to the QSA, but they are not used to compare the performance of students in one school with the performance of students in other schools. The scores are considered to be unsuitable for making comparisons across schools (see Queensland Studies Authority, 2010b, p. 19). Teachers make decisions about administration times (one, two, or more testing sessions) and when during the administration period to give the assessments, and they participate in the scoring process.

Assessment Task Example 11, Plate Tectonics: An example of a performance task that might be used for monitoring purposes is one that was administered in a classroom after students had covered major aspects of the earth and space science standards. It is taken from a program for middle school children in the United States that provided professional development based on *A Framework for K-12 Science Education: Practices, Crosscutting Concepts, and Core Ideas* (National Research Council, 2012a) and training in the use of curriculum materials aligned

---

[20]Low-stakes tests are those that do not directly affect a decision about any student or teacher.

to the framework. It was designed and tested as part of an evaluation of a set of curriculum materials and associated professional development.[21]

The task was given to middle school students studying a unit on plate tectonics and large-scale system interactions (similar to one of the disciplinary core ideas in the NGSS). The assessment targets two performance expectations linked to that disciplinary core idea. The task, part of a longer assessment designed to be completed in two class periods, is one of several designed to be given in the course of a unit of study. The task asks students to construct models of geologic processes to explain what happens over hot spots or at plate boundaries that leads to the formation of volcanoes. The students are given these instructions:

A. Draw a model of volcano formation at a hot spot using arrows to show movement in the model. Be sure to label all parts of your model.

B. Use your model to explain what happens with the plate and what happens at the hot spot when a volcano forms.

C. Draw a model to show the side view (crosssection) of volcano formation near a plate boundary (at a subduction zone or divergent boundary). Be sure to label all parts of your model.

D. Use your model to explain what happens when a volcano forms near a plate boundary.

In parts A and B of the task, students are expected to construct a model of a volcano forming over a hot spot using drawings and scientific labels, and they are to use this model to explain that hot spot volcanoes are formed when a plate moves over a stationary plume of magma or mantle material. In parts B and C, students are expected to construct a model of a volcano forming at a plate boundary using drawings and scientific labels and then use this model to explain volcano formation at either a subduction zone or divergent boundary.

The developers drew on research on learning progressions to articulate the constructs to be assessed. The team developed a construct map (a diagram of

---

[21]Although the task was designed as part of the evaluation, it is nevertheless an example of a way to assess students' proficiency with performance expectations like those in the NGSS. The question being addressed in the evaluation was whether the professional development is more effective when the curriculum materials are included than when they are not. Teachers in a "treatment" condition received professional development and materials needed to implement Project-Based Inquiry Science, a comprehensive, 3-year middle school science curriculum. The research team used evidence from the task discussed in this report, in combination with other evidence, to evaluate the integrated program of professional development and curriculum.

TABLE 5-2 Scoring Rubric for Task on Volcano Formation

Score Point	Descriptor B
+1	The explanation **states** or drawing clearly shows that *a volcano forms when magma from the hot spot rises and breaks through the crust.*
+1	The explanation **states** or drawing clearly shows that the *hot spot in the mantle stays in the same place and/or states that the crust/plate moves over it.*
0	Missing or response that cannot be interpreted.

thinking and understanding in a particular area; see Chapter 3) that identified disciplinary core ideas and key science practices targeted in the unit, which was based on research on how students learn about the dynamics of Earth's interior (Gobert, 2000, 2005; Gobert and Clement, 1999) and on research on learning progressions related to constructing and using models (Schwarz et al., 2009).

The scoring rubric in Table 5-2 shows how the task yields evidence related to the two performance expectations. (The developers noted that the task could also be used to generate evidence of student understanding of the crosscutting concepts of pattern and scale, although that aspect is not covered in this rubric.) The scoring rubric addressed the middle school performance expectations, as well as the range of student responses generated from a field test of the task. Field testing verified that students could provide explanations as part of their responses to the task that matched the researchers' expectations (Kennedy, 2012a,b).

Scores on the component sections of the task set were used to produce a single overall score (the individual parts of the item are not independent, so the task does not generate usable subscores). Taken together, the components demonstrate the "completeness" of a student's skill and knowledge in constructing models to explain how volcanoes form. To earn a top score for parts A and B, not only must students label key parts of their models (crust, plates, magma, and mantle) with arrows showing the mechanism involved, they must also provide an explanation of or clearly show how volcanoes form over a hot spot.

Figure 5-5 illustrates two students' different levels of performance on parts A and B. The drawing on the left received a combined score of 4 points (of a possible total of 5) for constructing a model because it includes labels for the mantle, magma, crust, volcano, and a hot spot. Arrows show the movement of crust, and the student has written a claim (below the drawing), "The hot spot allows magma to move up into the crust where it forms a volcano." The drawing includes the correct labels, shows some direction in the movement of the crust, and mentions

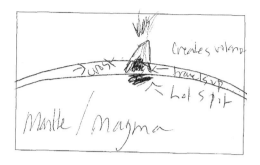

FIGURE 5-5 Two student responses to task on volcano formation.
SOURCE: SRI International (2013). Reprinted with permission.

magma moving up and penetrating the crust, to form a volcano. However, the student did not write or draw about the plate moving across the hot spot while the hot spot stays in the same place, so the model is incomplete.

The drawing on the right received only 1 point for parts A and B. It included a drawing of a volcano with magma and lava rising up, with the claim, "The magma pushes through the crust and goes up and erupts." The student's drawing does not show anything related to a hot spot, although it does mention that rising magma pushes up through the crust causing an eruption, for which the student earned partial credit.

A score on this task contributes one piece of evidence related to the performance expectations. A similar rubric is used to score parts C and D. These scores are combined with those on other tasks, given on other days, to provide evidence of student learning for the entire unit. No attempt is made to generate separate scores for the practice (developing models) and the knowledge because the model is a part of the way students are representing their knowledge in response to the task: these two aspects of practice and knowledge are not separable.

Portfolio of Work Samples and Projects

A third option for classroom-embedded assessments would be for a state or district to provide criteria and specifications for a set of performance tasks to be completed and assembled as work samples at set times during the year. The tasks might include assignments completed during a school day or homework assignments or both. The state or local school system would determine the scoring rubric and criteria for the work samples. Classroom teachers could be trained to score the samples, or the portfolios could be submitted to the district or state and scored centrally.

An alternative or complement to specifying a set of performance tasks as a work sample would be for a state or district to provide specifications for students to complete one or more projects. This approach is used internationally in Hong Kong; Queensland and Victoria, Australia; New Zealand; and Singapore. In these programs, the work project is a component of the examination system. The projects require students to investigate problems and design solutions, conduct research, analyze data, write extended papers, and deliver oral presentations describing their results. Some tasks also include collaboration among students in both the investigations and the presentations (Darling-Hammond et al., 2013).

Maintaining the Quality of Classroom-Embedded Components

The options described above for classroom administration as part of a monitoring assessment program introduce the possibility of local (district or school) control over certain aspects of the assessments, such as developing the assessments and involving teachers in administration or scoring the results. For these approaches to work in a monitoring context, procedures are needed to ensure that the assessments are developed, administered, and scored as intended and that they meet high-quality technical standards. If the results are to be used to make comparisons across classrooms, schools, or districts, strategies are needed to ensure that the assessments are conducted in a standardized way that supports such comparisons. Therefore, techniques for standardizing or auditing across classrooms, schools, and districts, as well as for auditing the quality of locally administered assessments, have to be part of the system.

Several models suggest possible ways to design quality control measures. One example is Kentucky's portfolio program for writing, in which the portfolios are used to provide documentation for the state's program review.[22] In Wyoming, starting officially in 2003, a "body of evidence system" was used in place of a more typical end-of-school exit exam. The state articulated design principles for the assessments and allowed districts to create the measures by which students would demonstrate their mastery of graduation requirements. The quality of the

---

[22]In Kentucky, the state-mandated program review is a systematic method of analyzing the components of an instructional program. In writing, the portfolios are used, not to generate student scores, but as part of an evaluation of classroom practices. For details, see http://education. ky.gov/curriculum/pgmrev/Pages/default.aspx [November 2013].

district-level assessments was monitored through a peer review process, using reviewers from all of the districts in the state (see National Research Council, 2003, pp. 30-32). Several research programs have explored "teacher moderation" methods. Moderation is a set of processes designed to ensure that assessment results (for the courses that are required for graduation or any other high-stakes decision) match the requirements of the syllabus. The aim of moderation is to ensure comparability; that is, that students who take the same subject in different schools or with different teachers and who attain the same standards through assessment programs on a common syllabus will be recognized as at the same level of achievement. This approach does not imply that two students who are recognized as at the same level of achievement have had exactly the same collection of experiences or have achieved equally in any one aspect of the course: rather, it means that they have on balance reached the same broad standards.

One example is the Berkeley Evaluation and Assessment Research Center, in which moderation is used not only as part of assessments of student understanding in science and mathematics, but also in the design of curriculum systems, educational programs, and teacher professional development.[23] Two international programs that use moderation, the Queensland program and the International Baccalaureate (IB) Program, are described in the rest of this section. The New Zealand Quality Assurance system provides another example.[24]

Example: Queensland Approach

Queensland uses a system referred to as "externally moderated school-based assessment" for its senior-level subject exams given in grades 11 and 12.[25] There are several essential components of the system:

- syllabuses that clearly describe the content and achievement standards,
- contextualized exemplar assessment instruments,
- samples of student work annotated to explain how they represent different standards,

---

[23]For details, see Draney and Wilson (2008), Wilson and Draney (2002), and Hoskens and Wilson (2001).

[24] For details, see http://www.k12center.org/rsc/pdf/s3_mackrell_%20new_zealand_ncea.pdf [November 2013].

[25]The description of the system in Queensland is drawn from two documents: *School-Based Assessment* (Queensland Studies Authority, 2010b) and *Developing the Enabling Contest for School-Based Assessment in Queensland, Australia* (Allen, 2012).

Developing Assessments for the Next Generation Science Standards

- consensus through teacher discussions on the quality of the assessment instruments and the standards of student work,
- professional development of teachers, and
- an organizational infrastructure encompassing an independent authority to oversee the system.

Assessment is determined in the classroom. School assessment programs include opportunities to determine the nature of students' learning and then provide appropriate feedback or intervention. This is referred to as "authentic pedagogy." In this practice, teachers do not teach and then hand over the assessment that "counts" to external experts to judge what the students have learned: rather, authentic pedagogy occurs when the act of teaching involves placing high-stakes judgments in the hands of the teachers.

The system requires a partnership between the QSA and the school. The QSA

- is set up by legislation;
- is independent from the government;
- is funded by government;
- provides students with certification;
- sets the curriculum framework (or syllabus) for each subject within which schools develop their courses of study;
- sets and operates procedures required to ensure sufficient comparability of subject results across the state; and
- designs, develops, and administers a test of generic skills (the Queensland Core Skills Test) with the primary purpose of generating information about groups of students (not individuals)

For each core subject (e.g., English, mathematics, the sciences, history):

- The central authority sets the curriculum framework.
- The school determines the details of the program of the study in this subject, including the intended program of assessment (the work program).
- The central authority approves the work program as meeting the requirements of the syllabus, including the assessment that will be used to determine the final result against standards defined in the syllabus.
- The school delivers the work program.

- The school provides to the central authority samples of its decision making about the levels of achievements for each of a small number of students on two occasions during the course (once in year 11 and once in year 12) with additional information, if required, at the end of year 12.
- Through its district and state panels, the central authority reviews the adequacy of the school's decision making about student levels of achievement on three occasions (once in year 11 and twice in year 12). Such reviews may lead to recommendations to the school for changes in its decisions.
- The central authority certifies students' achievement in a subject when it is satisfied that the standards required by the syllabus for that subject have been applied by the school to the work of students in that subject.

The QSA's task is to ensure, for example, that two students with the same result in a physics course from schools thousands of miles apart have met the same standards. Participating in consensus moderation meetings (or regional review panel meetings) is a core activity for teachers. In such meetings, they examine evidence about student performance from multiple schools, judge that evidence on the basis of the curricular standards, and give advice to schools about appropriate grades. Teachers secure significant professional recognition through participation in moderation panels.[26] Studies of this system indicate high levels of comparability and interrater agreement (Masters and McBryde, 1994; Queensland Studies Authority, 2010a). Over time, repeated participation in the moderation process provides professional development for teachers around critical issues of learning and of assessment.

Example: International Baccalaureate Program

Moderation procedures have also been used in the IB Program, which offers a diploma program worldwide for students ages 16 to 19.[27] The program includes both an internal assessment component, given by teachers, and standardized external assessment tests. The external assessments used for the sciences consist of three written paper-and-pencil tests that account for 76 percent of the final score. The internal assessment includes "an interdisciplinary project, a mixture of short- and

---

[26]For additional details about the moderation process, see Queensland Studies Authority (2010a). Available at http://www.qsa.qld.edu.au/downloads/senior/snr_moderation_handbook.pdf [November 2013].

[27]For details about the IB Diploma Program, see http://www.ibo.org/diploma/index.cfm [November 2013].

long-term investigations (such as laboratory and field projects and subject-specific projects) and, for design technology only, the design project" (International Baccalaureate Organization, 2007, p. 16). The internal assessment accounts for 24 percent of the final score.

The internal assessments are scored by teachers and externally moderated by the International Baccalaureate Organization (IBO). Grading of the internal assessments is based on assessment criteria published by the International Baccalaureate Organization (2007). For each criterion, there are descriptors that reflect different levels of achievement on student work products to guide grading. Teachers are required to submit a sample of candidates' work for moderation by external moderators (International Baccalaureate Organization, 2013). This is a two-step process in which (1) the moderator checks that the teacher applied the criteria provided for scoring of the internal assessment for a sample of students from different schools; and (2) the grades assigned by the teachers are adjusted by the IB Assessment Center whenever differences in interpretation or use of the criteria are identified (International Baccalaureate Organization, 2013, p. 94).

The grades assigned by the teacher may be raised, lowered, or left unchanged as a result of the moderation process. If the mean of candidates' moderated grades differ from the mean of the grades awarded by the teacher by 15 percent of the maximum possible score, a second moderation process is carried out. Grades may be raised as a consequence of the remoderation process, but they cannot be lowered (International Baccalaureate Organization, 2013, p. 72). Schools receive feedback on the suitability of the investigations they used as internal assessments and on the grades their teachers assigned, based on the assessment criteria. As in Queensland, this process is also regarded by the IB system as an essential component of teacher professionalism and professional development.

TAKING ADVANTAGE OF TECHNOLOGY

Our review of research and development of assessments designed for monitoring purposes, through either an on-demand or a classroom-embedded assessment component, has identified a number of important ways that both new and existing technologies can support the development of NGSS-aligned assessments. Mobile devices, computers, and other forms of technology can be used with any of the assessments we have described. Adapting assessments to technology-based enhancements opens up new possibilities for assessment tasks and for scoring and interpreting the results of tasks that assess three-dimensional science learning. Technology enhancements allow more opportunities for students to interact with tools, data, and the phenomena they are investigating (see, e.g., Pellegrino and Quellmalz, 2010-2011). Rich media (digital technology that allows for complex user interactions with a range of devices) has expanded the possibilities for simulations of situations that cannot easily be created in a classroom. Simulated investigations can be carried out quickly, allow multiple trials, and hence provide a tool to assess students' ability to plan and carry out investigations and to analyze data. New technology and platforms that support further upgrades make it much easier than in the past to accumulate, share, store, and transmit information. Such possibilities will make it easier to work with evidence collected in systems of assessment that are composed of multiple elements.

In addition, automated scoring is becoming more sophisticated and reliable, and new techniques are likely to become feasible—important developments because the scoring of open-ended questions can be labor intensive, time consuming, and expensive (see Nehm and Härtig, 2011). Scoring can take into account student actions and choices made in the course of an activity, as well as student responses to set tasks. All of these possibilities are likely to make it easier to assess constructs that are difficult to assess using paper-and-pencil tests. For example, using mathematics and computational thinking may be especially well suited to being assessed with technology.

However, there is a critical interplay between technology capability and task design: what the student can see and do with the technology and what actions or responses can be recorded. These elements can allow or deny particular aspects of tasks. In addition, care must be taken that all students being assessed have sufficient opportunity to familiarize themselves with the capabilities of the technology before being asked to use it in a testing situation. This is an important equity issue, as students from different backgrounds may have had very different levels of experience with such technologies both in and outside of their classrooms.

Technology also opens up new strategies for validly assessing students who are developing their language skills or who have other special needs, by making it easier to offer supports or accommodations and interfaces that use universal design principles to provide better access to the content of an assessment. Such universal design elements include audio reading of passages, translation of words or phrases, user-controlled pacing, varied size of text and volume, and multiple tools to reduce cognitive load and assist in organizing information.

Variations in Item-Response Formats

Technology expands the types of response formats that can be used in assessment tasks. Scalise and Gifford (Scalise, 2009, 2011; Scalise and Gifford, 2006) have developed a taxonomy that shows the variety of types of response formats that can be used in tasks presented on the computer. Figure 5-6 shows an "intermediate constraint taxonomy"[28] that categorizes 28 innovative item types that can be used with computer-based assessments. Item-response formats range from fully constrained (such as the conventional multiple-choice format shown in cell 1C) to fully unconstrained (such as the traditional essay shown in cell 6D). Intermediate constraint items are more open ended than fully multiple-choice formats, but they allow students to respond in a way that is machine scorable.

For instance, option 4A in the taxonomy is referred to as an interlinear option. This is an alternative to a traditional fill-in-the-blank format. With this format, a student is presented with a brief written passage that contains a few blanks. Using technology, a set of choices is offered for each blank, and the student clicks on his or her choice. Option 4B is referred to as the sore finger option: the student is presented with a model and asked to identify the incorrect part by placing an X on the incorrect piece of the model. Thus, the question does not simply offer a set of options of models for the student to choose from (as would be the case in a multiple-choice format), nor does it require the student to draw the model from scratch.

Other cells of the taxonomy represent additional options. Option 3B, categorizing, is a format that allows students to drag-and-drop items so that they are properly classified. The ranking and sequencing option in 3C asks students to put a series of events in proper order. The various item-response formats shown in the table provide a variety of alternatives to the traditional multiple-choice and fully open-response formats. Technology is a crucial component of a number of these response formats.

Example: Technology-Enhanced Version of an AP Biology Question

Using the options in the above taxonomy—or other approaches to innovative formats—technology-enhanced assessments can be designed to address particular assessment challenges. Using assessment design approaches that draw on

---

[28]For an interactive version of the taxonomy, see http://pages.uoregon.edu/kscalise/taxonomy/taxonomy.html [June 2013].

1. Multiple Choice	2. Selection/ Identification	3. Reordering/ Rearrangement	4. Substitution/ Correction	5. Completion	6. Construction
1A. True/False	2A. Multiple True/False	3A. Matching	4A. Interlinear	5A. Single Numerical Constructed	6A. Open-Ended Multiple Choice
1B. Alternate Choice	2B. Yes/No with Explanation	3B. Categorizing	4B. Sore-Finger	5B. Short-Answer and Sentence Completion	6B. Figural Constructed Response
1C. Conventional Multiple Choice	2C. Multiple Answer	3C. Ranking and Sequencing	4C. Limited Figural Drawing	5C. Cloze-Procedure	6C. Concept Map
1D. Multiple Choice with New Media Distractors	2D. Complex Multiple Choice	3D. Assembling Proof	4D. Bug/Fault Correction	5D. Matrix Completion	6D. Essay and Automated Editing

FIGURE 5-6 The intermediate constraint taxonomy, a categorization of 28 innovative item types useful in computer-based assessment.

SOURCE: Scalise (2009). Reprinted with permission.

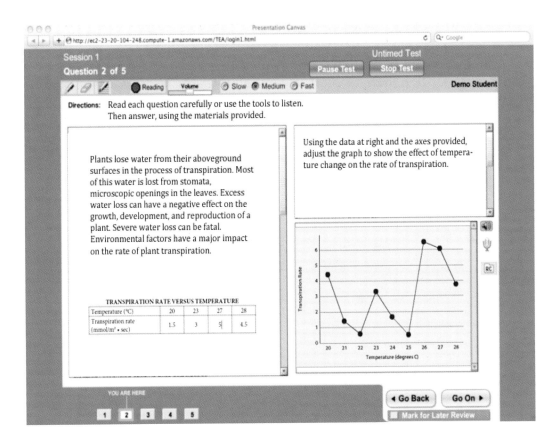

FIGURE 5-7 Original example from the AP biology assessment.
SOURCE: Adapted from Barton and Shultz (2012) and College Board (2012).

strong evidentiary reasoning (see Chapter 3), a task can be created using the new formats and put in an appropriate delivery environment. A hypothetical example of this is shown in Figures 5-7 through 5-10.[29] The task was originally designed for a paper-and-pencil format, as shown in Figure 5-7. In the next three figures, we have adapted it for use in a technology-enhanced environment. The delivery environment shown here, into which the example task has been integrated, is drawn from an example presented recently for assessing hard-to-measure constructs in the Common Core State Standards (Barton and Schultz, 2012).

---

[29]This example was adapted from an AP biology task in the preparatory materials for students and is available at http://media.collegeboard.com/digitalServices/pdf/ap/IN120084785_ BiologyCED_Effective_Fall_2012_Revised_lkd.pdf [December 2013].

Plant Respiration

The following table shows transpiration at different temperatures.

- Drag the points to best show the effect of temperature change on the rate of transpiration.

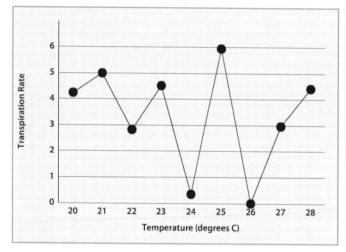

Year	Trout Population
20	1.5
21	not available
22	not available
23	3
24	not available
25	not available
26	not available
27	5
28	4.5

FIGURE 5-8 AP biology example placed into a technology-enhanced format.
NOTE: Each student receives a new version of the graph from Figure 5-7.
SOURCE: Adapted from Barton and Shultz (2012) and College Board (2012).

In this example, an interactive graph has been created that the student adjusts to answer the question. The format (type 6A—see Figure 5-6, above) is open ended multiple-choice. Rather than having only a few choices, as in a traditional multiple-choice format, in this format all or a large portion of the possible outcome space is available for the student. In other words, by sliding the points on the display to any location, students create their own version of the graph, similar to a constructed response on paper. A student's complete graph is shown in Figure 5-10.

This format contrasts with the selected-response format used for traditional multiple-choice questions, in which perhaps four or five versions of a graph are provided from which the student would select the graph display that best answers the question. The problem with that format is that when only a few options are shown, students can "backsolve": that is, instead of directly solving the problem, they can test each of the provided solutions. Furthermore, when a limited range of answer choices is provided, student thinking may be prompted by the visual displays provided in the

Plant Respiration

The following table shows transpiration at different temperatures.

• Drag the points to best show the effect of temperature change on the rate of transpiration.

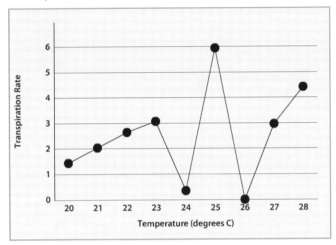

Year	Trout Population
20	1.5
21	not available
22	not available
23	3
24	not available
25	not available
26	not available
27	5
28	4.5

FIGURE 5-9 First graph adjustment, using drag-and-drop procedures.

NOTE: The student has adjusted the first four points (for temperatures of 20 to 23 degrees).

SOURCE: Adapted from Barton and Shultz (2012) and College Board (2012).

choices. In such cases, understanding of a complex concept may be less well measured due to the "short-cut" paths to a solution suggested by the small set of possible answers that are provided. Open ended multiple choice, by contrast, is still a type of selection—students select points and move them to new positions—but the prompting and possibilities for backsolving are reduced by not displaying answer choices. Furthermore, as an intermediate constraint format, it is readily scorable by computer. Also, task variants with unique starting points for the display, for instance, can easily be created.

Example: Technology-Enhanced Tasks on NAEP

Another example of the ways in which technology enhancements can be used is provided by the 2009 NAEP Interactive Computer and Hands-On Tasks Science Assessment. This assessment, given to national samples of students in the 4th, 8th, and 12th grades, was designed to produce national results for each grade. For each grade level, each student is assigned three computer-interactive tasks, two

Plant Respiration

The following table shows transpiration at different temperatures.

- Drag the points to best show the effect of temperature change on the rate of transpiration.

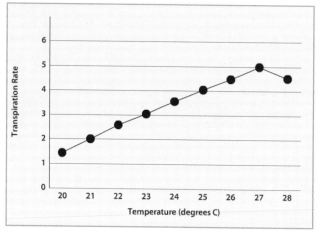

Year	Trout Population
20	1.5
21	not available
22	not available
23	3
24	not available
25	not available
26	not available
27	5
28	4.5

FIGURE 5-10 Complete graph for all temperatures in the data table.

NOTE: When finished, the points should reflect the most likely graph, given the points in the data table.

SOURCE: Adapted from Barton and Shultz (2012) and College Board (2012).

intended to take 20 minutes to answer and one designed to take 40 minutes.[30] The tasks included a variety of types of simulations through which students follow instructions for designing and carrying out experiments and recording and making graphs of the data. The tasks make use of a variety of response formats, including multiple choice, short answers, and drag-and-drop procedures.

For example, in one of the 4th-grade tasks, students were asked to investigate the effects of the temperature changes on a concrete sidewalk.[31] The simulation first presented students with a flask of water and asked them to observe and record what happens to the volume when the temperature is raised and lowered so that the water melts and then freezes. In completing the task, the students are asked to make observations, develop explanations that they support with

---

[30]All the tasks are publicly available at http://nationsreportcard.gov/science_2009 [June 2013].
[31]The task is available at http://nationsreportcard.gov/science2009ict/concrete/concrete1.aspx [June 2013].

evidence, and then use the simulation to predict what will happen to cracks in concrete when the temperature increases and decreases. The students complete the task by generating a written remedy for preventing further cracking of the concrete.

One of the 8th-grade tasks asked students to evaluate the environmental effects associated with developing a new recreation area.[32] This task began by presenting information about three types of environments—forest, wetland, and meadow—that are being considered for the recreation area and about eight animals that reside in these environments. A simulation is used to take students through the relationships in a food web, prompting them with questions about the animals' eating habits to ensure the students understand the concept of a food web. The simulation then asks students to use the information from the food web to explain or predict what would happen if the population of certain animals decreased and to apply that information to the problem of evaluating the environmental effects of locating the recreation area in each of the three environments. This part of the simulation takes students through the task of creating and explaining a set of graphs. The task concludes by asking students to write a recommendation for the location of the recreation area, justify the recommendation with evidence, and discuss the environmental effects.

It is important to point out that although these tasks do involve new ways of assessing science learning, they were not designed to measure the type of three-dimensional science learning that is in the NGSS. But they do demonstrate some of the capabilities in large-scale assessment that become possible with simulations and other technological approaches. The 2009 NAEP assessment moved the field substantially forward, but as noted in *Leading Assessment into the Future*,[33] the report on the NAEP assessment, there is much work still needed in this field.

Assessing Challenging Constructs

Technology can also make more evidence available for hard-to-measure constructs, such as demonstrating proficiency in planning and carrying out investigations, through the use of simulations, animations, video, and external resources with scientific data and results.

---

[32]The task is available at http://nationsreportcard.gov/science2009ict/park/park.aspx [June 2013].

[33]See nces.ed.gov/.../FutureOfNAEP_PanelWhitePaperFINAL05.15.2012.pdf [December 2013].

ARCTIC TREK

Collaboration contest

For this collaboration contest, you work with your team and use clues to discover a series of 6 answers.

HINT:

Here is how a clue works. The first part of the clue directs you to one of the web sites listed to the right. The rest of the clue guides you through the site to find the answer.

This is a timed contest to see what team can come up with the 6 answers first. Good Luck and Happy Hunting!

FIGURE 5-11 Introduction to the polar bear task.
SOURCE: Wilson et al. (2012). Copyright by the author; used with permission.

An example of a task that makes use of innovative technologies is provided by an assessment module called the Arctic Trek scenario developed by Wilson and colleagues (2012) for the Assessment and Teaching of 21st Century Skills (ATC21S) project.[34] For this module, students work in teams to respond to questions about polar bear populations in the world. The module provides access to various pages, and the student teams are to determine which webpages provide the information needed to respond to the questions. The teams assign themselves roles in responding to the tasks (e.g., captain, recorder), and the technology allows them to chat with each other as they gather information to answer questions and complete a notebook.

Figure 5-11 shows a screenshot that introduces the module to the students. An example of a question from the module is shown in Figure 5-12. The student

---

[34]See http://atc21s.org for information about this project. To see additional details about the example task, see http://atc21s.org/wp-content/uploads/2012/12/white-paper6-Assessment-of-Learning-in-Digital-Social-Networks_DRAFT.pdf [December 2013].

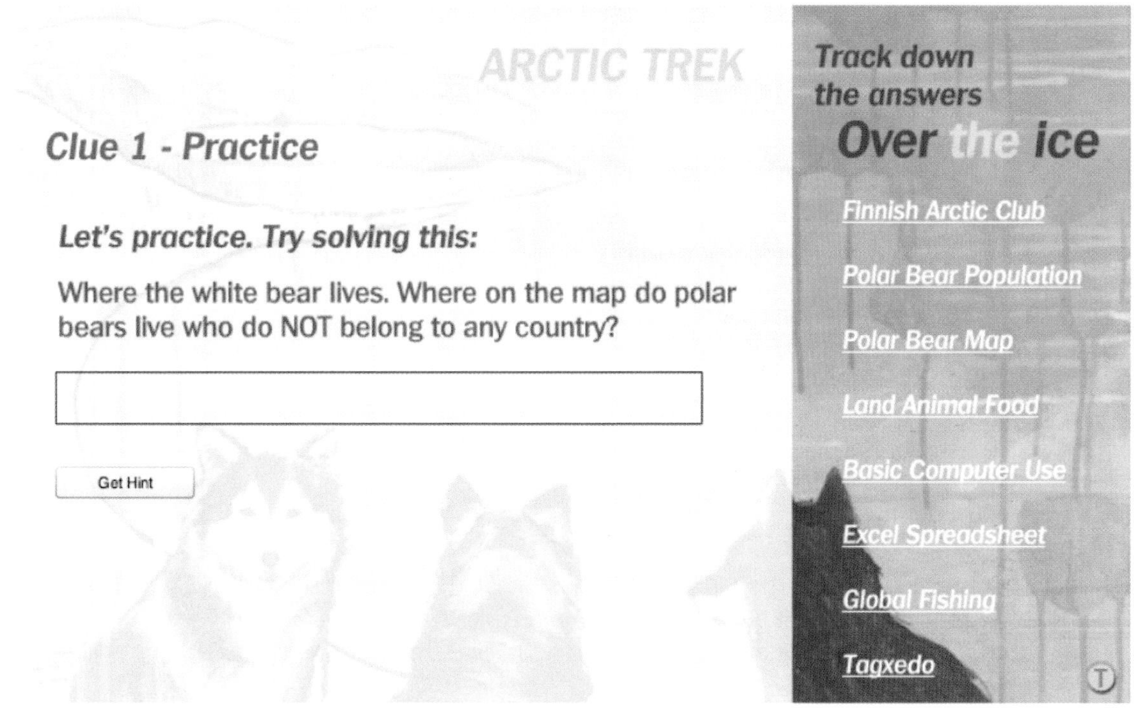

FIGURE 5-12 Example of a question for the polar bear task.

NOTE: To answer the question, the student must determine which of the websites (from the list on the right) will provide the needed information, click on the website needed to answer the question, and find the needed information. This "clue" is used for practice to familiarize students with the technology.

SOURCE: Wilson et al. (2012). Copyright by the author; used with permission.

is expected to select the appropriate website to answer the question "Where do polar bears live that do not belong to any country?" In this case, the question is designed for practice to acquaint the student with the technology. If a student does not know how to the answer the question, the student can request a hint (and this can be repeated). Figure 5-13 shows the question with a hint. If the hints are not enough (and eventually they end up telling the student exactly what to do), then the student may request teacher assistance by hitting the "T" button, which appears at the bottom right-hand corner of the list of websites. The software allows the teacher to track students' work, and in this case, the teacher is to fill in a box with information that can be used as part of the scoring: see Figure 5-14.

Clue 1 - Practice

Let's practice. Try solving this:

Where the white bear lives. Where on the map do polar bears live who do NOT belong to any country?

[   ]

[Another Hint]

The first sentence of the clue helps you select a webpage from the list at right. Which page is about where white bears (polar bears) live? Click on that link and find a map. Use the map to answer the question.

FIGURE 5-13 A hint to guide the student in selecting the correct link for the polar bear task.
SOURCE: Wilson et al. (2012). Copyright by the author; used with permission.

An actual task is shown in Figure 5-15, in which a student would read through an online display. Here the student has been asked to examine a map that shows where polar bears are found and must describe the way information is conveyed on the map. Each student responds to this task individually and then shares her or his response with the team.

The technology also allows the teacher to track their interactions and responses and to provide assistance when needed. Although this task is designed to measure social interaction and teamwork, the approach could easily be adapted to allow students to demonstrate their proficiency with various scientific and engineering practices. The module is designed for group work, with close monitoring by the teacher, but it could easily be adapted to be used for summative assessment purposes.

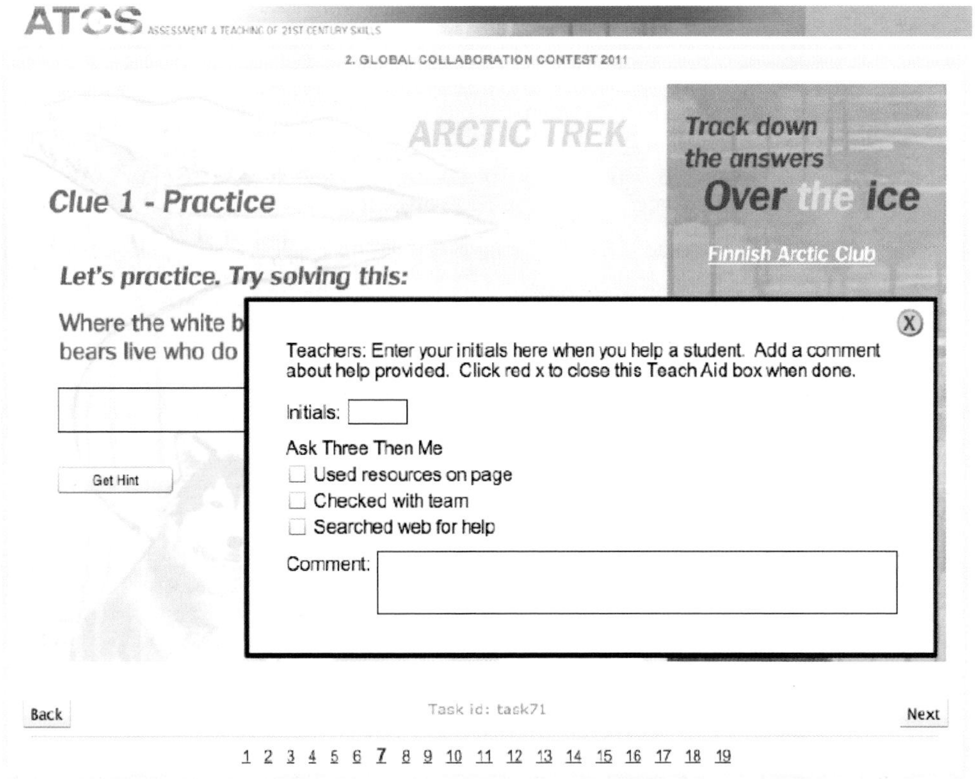

FIGURE 5-14 Information box for a teacher to record the level of assistance a student required for the polar bear task.
SOURCE: Wilson et al. (2012). Copyright by the author; used by permission.

Task Surrounds

In the context of technology-enhanced assessment, a task surround is a set of small software programs that work together to create a set of activities, such as for a research or inquiry activity, which can be readily populated with new content (Scalise, 2011, p. 8). A task surround can be used to develop additional tasks that all use the same technology. Once developed, a task surround ("shell") can be used repeatedly with a range of new content and different tasks, making the investment in the technology more affordable and the technology itself more familiar to students.

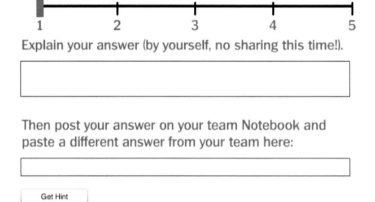

FIGURE 5-15 Example of an actual question from the polar bear task.
SOURCE: Wilson et al. (2012). Copyright by the author; used with permission.

A task surround provides a computer-based, hands-on, or remote lab instructional platform with common interfaces for a variety of routine tasks, such as running simulations, graphing results, viewing animations, and consulting reference materials and links (Buckley et al., 2004; Gobert and Pallant, 2004; Gobert et al., 2003). A surround is more than a basic interface in that it can be changed to represent different standards and domains or to produce a range of task variants within a standard or domain. The task surround can be varied in the range of functionalities provided from one task to the next to fit different design patterns (see the pinball car example in Chapter 3), constructs, or goals and objectives of measurements. When the task surround incorporates new content intended to address the same goals and objectives of the original content, it is called a task variant. Task variants can be used to develop alternative forms of an assessment. When the task surround is populated with prompts and materials intended to be quite different from the original version, it is an example of technology-enhanced generalization. Reuse of a surround can serve many different purposes: each purpose can use the same programming and technology investment.

For large-scale assessment, numerous technology-enhanced approaches built around interactive scenarios, reusable components, and task surrounds are emerging. These have been used in several recent assessments, including the OECD's 2012 PISA (Steinhauer and Koster Van Goos, 2013), the 2009 NAEP (described above), the 2013 *International Computer and Information Literacy Study of the International Association for the Evaluation of Educational Achievement* (International Association for the Evaluation of Educational Achievement, 2013*)*, and even to some extent in the interactive activities being designed and piloted by the U.S. Race to the Top Assessment Program consortia (see Chapter 1). The pinball car example discussed in Chapter 3 provides an example of a task surround. The design pattern (see Figure 3-3) lays out the key elements for the task and could be used to generate a number of different tasks that use the same technology, software, or both.

CONCLUSIONS AND RECOMMENDATIONS

Our review of various strategies for administering assessments of three-dimensional science learning in formats that will yield results to support system monitoring makes clear that there are tradeoffs with a number of competing goals. One goal is to use assessments composed principally of performance tasks, particularly those that allow students to actually demonstrate their skills using hands-on tasks. But another goal is to minimize the amount of time students spend on assessment in order to leave more time for instruction. Yet another goal is to have assessments that produce scores that are sufficiently reliable and valid to support high-stakes uses and sufficiently comparable to provide information about cross-group and cross-time comparisons, such as to answer the questions in Table 5-1 (above). Still another goal is to achieve the desired assessment at a reasonable cost level relative to the intended measurement benefits.

The measurement field has progressed considerably since the 1990s when performance tasks and portfolios were last tried on a large scale. Much has been learned from those prior attempts, and more possibilities are now available with technology. More is known about ways to develop tasks, standardize the way that they are administered, and score them accurately and reliably. In addition, the field now acknowledges that reliability statistics for individual-level scores and decisions are different from those for higher levels of aggregations, for example, at the school or district level. Technological innovations provide platforms for presenting tasks in more realistic ways, measuring constructs that could not previously be measured, incorporating features to make tasks more accessible to all students, and administering and scoring performance-based tasks and portfolios more efficiently.

Nevertheless, a number of challenges remain. As noted above, it will not be possible to cover all of the performance expectations for a given grade in one testing session. Even with multiple testing sessions, external on-demand assessments alone will not be sufficient to fully assess the breadth and depth of the performance expectations.

CONCLUSION 5-1 To monitor science learning and adequately cover the breadth and depth of the performance expectations in the Next Generation Science Standards, information from external on-demand assessments will need to be supplemented with information gathered from classroom-embedded assessments. These assessments will need to be designed so that they produce information that is appropriate and valid to support a specific monitoring purpose.

The use of classroom-embedded assessments means that some of the testing decisions will have to be made locally by schools or districts. Those decisions include the timing and conditions of the administration and, possibly, the scoring procedures. These procedures will need to be carefully monitored to ensure that they are implemented as intended and produce high-quality data.

CONCLUSION 5-2 When classroom-embedded assessments are used for monitoring purposes, quality control procedures will be needed to ensure that assessments are administered and scored as intended and the data they produce are of high quality.

In the past decade, matrix sampling has not been widely used on external assessments used for monitoring purposes because of the intense focus on individual student scores under NCLB. However, it can be a useful and powerful tool in developing assessments of the NGSS and to meet certain monitoring purposes.

CONCLUSION 5-3 Matrix sampling will be an important tool in the design of assessments for monitoring purposes to ensure that there is proper coverage of the full breadth and depth of the NGSS performance expectations.

The approaches we propose for designing monitoring assessments that include performance tasks and portfolios may not yield the level of comparability of results that educators, policy makers, researchers, and others have been accustomed to, particularly at the individual student level. In proposing these approaches, we made the assumption that developing assessments that validly measure students' proficiency on the NGSS is more important than achieving strict comparability. However, we also think that focused research on strategies for enhancing the comparability of results from the approaches we propose will yield improvements in this area.

RECOMMENDATION 5-1 Research will be needed to explore strategies for enhancing the comparability of results from performance tasks and portfolio assessments of three-dimensional science learning so that they yield results that are appropriate for the intended monitoring purpose. Appropriate use of such strategies will need to include acceptance of alternative concepts and varying degrees of comparability among assessments according to their usage. Specifically needed is research on methods for statistically equating and/or linking scores and on methods for using moderation techniques. Such research should build on the existing literature base of prior and current efforts to enhance the comparability of scores for these types of assessments, including studies of approaches used in other countries.

Innovations in technology and in assessment design hold promise for addressing some of the challenges associated with the assessment approaches we suggest and should be considered to the extent that they produce valid and reliable outcomes.

RECOMMENDATION 5-2 Assessment developers should take advantage of emerging and validated innovations in assessment design, scoring, and reporting to create and implement assessments of three-dimensional science learning. To the extent that they facilitate achieving valid and reliable outcomes, available technological approaches should be used in designing, administering, and scoring science assessments.

As the field moves forward with these innovations, it will be important to verify that they meet the necessary technical standards.

RECOMMENDATION 5-3 Assessment developers and researchers should thoroughly evaluate the technical quality of science assessments used for monitoring purposes to verify that they meet the technical and validity standards required for their intended purpose.

6

DESIGNING AN ASSESSMENT SYSTEM

In this chapter, we turn to the question of how to design a full assessment system and consider the components that should be included to adequately evaluate students' science achievement. The assessment system we envision builds on discussion in the previous chapters of the report.

Chapter 2 explores the assessment challenges associated with evaluating students' proficiency on the performance expectations of the Next Generation Science Standards (NGSS) and emphasizes that because of the breadth and depth of those expectations, students will need multiple opportunities to demonstrate their proficiencies using a variety of assessment formats and strategies. The chapter also discusses the types of tasks that are best suited to assessing students' application of scientific and engineering practices in the context of disciplinary core ideas and crosscutting concepts, as well as simultaneously assessing the connections across concepts and disciplines. The committee concludes that tasks composed of multiple interrelated questions would best serve this purpose. Chapter 3 describes approaches to developing these types of tasks so that they provide evidence to support the desired inference. Chapters 4 and 5 present examples and discuss strategies for developing assessments for use, respectively, in the classroom and to provide evidence for monitoring purposes.

We propose that an assessment system should be composed both of assessments designed to support classroom teaching and learning (Chapter 4) and those designed for monitoring purposes (Chapter 5). In addition, the system should include a series of indicators to monitor that the students are provided with adequate opportunity to learn science in the ways laid out in *A Framework for K-12 Science Education: Practices,*

Crosscutting Concepts, and Core Ideas (National Research Council, 2012a, Chapter 11, hereafter referred to as "the framework") and the *Next Generation Science Standards: For States, By States* (NGSS Lead States, 2013, Appendix D). Such a system might take various forms and would include a range of assessment tools that have each been designed and validated to serve specific purposes and to minimize unintended negative consequences. Our intention is not to prescribe a single design for such a system, but to offer guidance for ensuring that any given system design supports attainment of the framework's vision for science learning and student proficiency envisioned in the framework and the NGSS.

We begin with the rationale for a systems approach to assessment, describing how an assessment system influences student learning and curriculum and instruction directly and indirectly and discussing the influence that accountability goals can have on the design of an assessment system. In the last section we describe a set of components and the characteristics that an effective assessment system should have and recommend strategies for developing such a system.

RATIONALE FOR A SYSTEMS APPROACH

As discussed throughout this report, the purposes for which information about student learning is needed should govern the design and use of assessments. These purposes may include

- guiding and informing teachers' day-to-day instructional decisions;
- providing feedback to students, as well as their parents and teachers, on students' academic progress;
- illustrating sound instructional and assessment activities that are consistent with the framework and the NGSS;
- monitoring the science achievement of students across schools, districts, states, and/or the nation to inform resource allocations, identify exemplary practices, and guide educational policy;
- contributing to the valid evaluation of teachers, principals, and schools;
- determining whether students meet the requirements for a high school diploma; and
- evaluating the effectiveness of specific programs (e.g., new science curricula and professional development to support the transition to the NGSS).

Implicit in each assessment purpose are one or more mechanisms through which the assessment is intended to have some beneficial effect. That is, assessments are a means to an end, not an end themselves. For example, an assessment that periodically informs students and their parents about student progress might be intended to stimulate students' motivation so that they study more, provide feedback students can use to focus their studies on their weak areas, and engage parents in student learning so they can provide appropriate supports when students are not making the expected level of progress. Similarly, providing teachers with quick-turnaround feedback on student learning might be intended to help them pace instruction in an optimal way, highlight individual learning difficulties so they can provide individualized remediation, and guide ongoing refinement of curriculum and instructional practices. Assessments that provide overall information about student learning might be used to evaluate the quality of instruction at the school, district, or state level in order to determine where to focus policy interventions. Assessments used for accountability purposes may be designed to hold teachers or schools and their principals accountable for ensuring that students achieve the specified level of progress.

Some of these action mechanisms are direct, in that the information provided by the test scores is used to inform decisions, such as to guide instruction or to make decisions about student placement. Other mechanisms are indirect, in that the testing is intended to influence the behavior of students, teachers, or educational administrators by providing them with incentives to improve test performance (hence, achievement). Assessments can provide teachers and administrators with examples of effective assessment practices that can be incorporated into instruction. Systems that involve teachers in the assessment design and scoring process provide them with an opportunity to learn about the ways students learn certain concepts or practices and about the principles and practices of valid assessment. Similarly, students who must pass an examination to receive a high school diploma may work harder to learn the content to be tested than they would without that requirement. Elementary grade teachers might invest more time and effort in science teaching if science test results were among the factors considered in accountability policies.[1] Other action mechanisms are even more indirect. For example, the content and format of testing send signals to textbook writers, teachers, and students about what it is important to learn (Haertel, 2013; Ho, 2013). Test questions that are made public and media reports of student results may help educate both educational professionals and the broader public about science learning and its importance.

---

[1]We acknowledge that both of these uses are controversial.

Value of a System of Assessments

Clearly, no single assessment could possibly serve the broad array of purposes listed above. Different assessment purposes require different kinds of assessment data, at different levels of detail, and produced with different frequency. Teachers and students, for example, need fine-grained, ongoing information unique to their classroom contexts to inform immediate instructional decision making; policy makers need more generalized data both on student learning outcomes and on students' opportunities to learn.

The arguments for the value of an assessment system have been made before (e.g., National Research Council, 2001). A systems approach to science assessment was advocated and described in considerable detail in *Systems for State Science Assessment* (National Research Council, 2005) and is reinforced in the new framework (National Research Council, 2012a). More recently, a systems approach was recommended in connection with the implementation of the Common Core State Standards in *Criteria for High-Quality Assessments* (Darling-Hammond et al., 2013). These reports all call for a balanced, integrated, and coherent system in which varied assessment strategies, each intended to answer different kinds of questions and provide different degrees of specificity, produce results that complement one another. In particular, the framework makes clear that an effective system of science assessment will include both assessments that are grounded in the classroom and assessments that provide information about the effectiveness of instruction and the overall progress of students' science learning.

The challenges of covering the breadth and the depth of the NGSS performance expectations amplify the need for a systems approach. The selection and design of system components should consider the constructs and purpose(s) each measure is to serve and the ways in which the various measures and components will operate to support the improvement of student learning. There are many ways to design an effective assessment system, but all should begin with careful consideration of the way that the assessment data are to be used, the type of information that is needed to support those uses (in the shape of a menu of different types of reports), and how the various components of the system work together.

Curriculum and Instruction

It important to point out that no assessment system operates in a vacuum. As argued in previous reports (National Research Council, 2001, 2005; Darling-Hammond et al., 2013), an assessment system should be designed to be coherent with instruction and curriculum.[2] The committee believes that curriculum design decisions should precede assessment design decisions. That is, decisions about which practices, crosscutting concepts, and core ideas will be taught need to be made before one can determine what will be assessed and how it will be assessed.

The NGSS illustrate an extensive set of performance expectations at every grade level. As we note in Chapter 2, it is unrealistic to suppose that each possible combination of the three dimensions will be addressed. Thus, in designing curricula, difficult decisions will have to be made and priorities set for what content to teach and assess.

In the United States, curricular decisions are made differently in each state: in some states, these decisions are made at the state level; in others, they are made at the district level or school level. Although the NGSS imply certain approaches toward curriculum design, education policy makers in different jurisdictions will make different decisions about what is the optimal curriculum for addressing the framework. Different curricula will likely reflect different priorities and different decisions about what to include.

These state differences pose a challenge for external assessments[3] when the assessment purpose is to compare performance across different jurisdictions, such as across states that have adopted different curricula or across schools and districts in states with local control over curricula. When external assessments are used to make comparisons, they will need to be designed to be valid, reliable, and fair despite the fact that students have been exposed to different curricula and different combinations of scientific practices, crosscutting concepts, and disciplinary core ideas. Students who have been exposed to any curriculum that is intended to be aligned with the framework and the NGSS should be able to show what they know and can do on assessments intended to support comparative judgments.

Devising assessments that can produce comparable scores that reflect complex learning outcomes for students who have studied different curricula is always a challenge. Test content needs to be neither too unfamiliar nor too familiar if it is to measure the intended achievement constructs. The challenge is to limit and balance the ways in which curriculum exposure may bias the results of an assessment that is to be used to make comparisons across student groups. These challenges in assessment design are not unique to science assessment. Test devel-

---

[2]"Curriculum" refers to the particular material through which students learn about scientific practices, crosscutting concepts, and core ideas.

[3]We use the term to mean assessments developed outside of the classroom, such as by the state or the district. External assessments are generally used for monitoring purposes (see Chapter 5).

opers in the United States have long had to deal with the challenge of developing external assessments that are fair, reliable, and valid for students who have studied different curricula. However, covering the full breadth and depth of the NGSS performance expectations is an additional challenge and will require a careful and methodical approach to assessment design.

Accountability Policies

The science assessments developed to measure proficiency on the NGSS performance expectations will likely be used for accountability purposes, so it is important to consider the ways in which accountability policies might affect the ways in which the assessments operate within the system. The incentives that come with accountability can serve to support or undermine the goals of improving student learning (National Research Council, 2011b; Koretz, 2008). It is likely that whoever is held accountable in a school system will make achieving higher test scores a major goal of science teaching.

In practice, accountability policies often result in "teaching to the test," so that testing tends to drive curriculum and instruction, even though the avowed intention may be for curriculum and instruction to drive testing (Koretz, 2005, 2008). The result of accountability testing, too often, has been a narrowing of the curriculum to match the content and format of what is to be tested, which has led to coverage of superficial knowledge at the expense of understanding and inquiry practices that are not assessed (Dee et al., 2013). Schools and classrooms serving students with the greatest educational needs are often those presented with the most ambitious challenges for improvement and thus also face the greatest pressure to "teach to the test." Thus, it is extremely important that the tests used for accountability purposes measure the learning that is most valuable.

As we have discussed in Chapters 4 and 5, the three-dimensional learning described in the framework and the NGSS cannot be well assessed without some use of the more extended engagements that are really only possible in a classroom environment. We emphasize that the assessments used for monitoring purposes will need to include both on-demand and classroom-embedded assessment components (see Chapter 5).[4] Thus, if accountability policies are part of the

---

[4]These two types of assessments were discussed in Chapter 5. We use them to mean the following. On-demand assessments are external assessments mandated by the state (such as the statewide large-scale assessments currently in place). They are developed and/or selected by the state and given at a time determined by the state. Classroom-embedded assessments are external assessments developed and/or selected by the state or the district. They are given at a time

science education system, it will be important that they incorporate results from a variety of types of assessments. When external, on-demand assessments dominate in an assessment system and are the sole basis for accountability, curriculum and instruction are most likely to become narrowed to reflect only the material and testing formats that are represented on those assessments (Koretz, 2005, 2008).

There is very limited evidence that accountability policies to date, which focus largely—if not solely—on external (large-scale) assessments, have led to improved student achievement (National Research Council, 2011b). In contrast, numerous studies document the positive effects on learning from the use of classroom assessment to guide teaching and learning (Black and Wiliam, 1998; Kingston and Nash, 2011; National Research Council, 2007). Assessment that closely aligns with a curriculum that engages students in three-dimensional science learning will return the focus to what is most important—the direct support of students' learning.

Communicating Assessment Results

A key consideration in developing an assessment system is the design of reports of assessment results. The reporting of assessment results is frequently taken for granted, but consideration of this step is critical. Information about students' progress is needed at all levels of the system. Parents, teachers, school and district administrators, policy makers, the public, and students need clear, accessible, and timely information. In a systems approach, many different kinds of information need to be available, but not all audiences need the same information. Thus questions about how various kinds of results will be combined and reported to different audiences and how reporting can support sound, valid interpretations of results need to be considered early in the process of the design of an assessment system.

Reporting of assessment results can take many forms—from graphical displays to descriptive text and from a series of numbers to detailed analysis of what the numbers mean. Depending on the needs of different audiences, results can be presented in terms of individual standards (or performance expectations) or in terms of clusters of standards. Results can describe the extent to which students have met established criteria for performance, and samples of student work can be provided.

---

determined by the district or school. See Chapter 5 for additional details about our use of these terms.

The types of assessments we advocate will generate new kinds of information. If the information is not presented in a way that is accessible and easy to use for those who need it, it will not serve its intended purpose. For example, if a series of complex performance tasks results in a single reported score, users will not be receiving the information the assessment was designed to produce. Thus, it is important that the reporting of assessment results be designed to meet the needs of the intended audiences and the decisions they face and address all of the specifications that guided the design and development of the assessment. For example, to be useful to teachers, assessment results should address instructional needs. Assessment reports should be linked to the primary goals of the framework and the NGSS so that users can readily see how the specific results support intended inferences about important goals for student learning. It is also important that the information provide clear guidance about the degree of uncertainty associated with the reported results.

The topic of developing reports of assessment results has been explored by a number of researchers: see, for example, Deng and Yoo (2009); Goodman and Hambleton (2004); Hambleton and Slater (1997); Jaeger (1996); National Research Council (2006); Wainer (2003).

SYSTEM COMPONENTS

The committee concludes that a science assessment system should include three components: (1) assessments designed for use in the classroom as part of day-to-day instruction, (2) assessments designed for monitoring purposes that include both on-demand and classroom-embedded components, and (3) a set of indicators designed to monitor the quality of instruction to ensure that students have the opportunity to learn science as envisioned in the framework. The first two components are only briefly considered below since they are the focus of extended discussion in Chapters 4 and 5. We emphasize below the third component—a set of indicators of opportunity to learn.

The approach to science assessment that we envision is different from those that are now commonly used (although it is indeed an extension and coordination of aspects of many current assessment systems). For instance, classroom-generated assessment information has not been used for monitoring science learning in the United States. Adopting an assessment system that includes a classroom-embedded component will require a change in the culture of assessment, particularly in the level of responsibility entrusted to teachers to plan, implement, and score assessments. In Chapter 5, we discuss ways to enhance the comparability of assessment information gathered at the local level by using moderation strategies[5] and methods for conducting audits to ensure that the information is of high quality. In addition, it will be important to routinely collect information to document the quality of classroom instruction in science, to monitor that students have had the opportunity to learn science in the way called for in the new framework, and to ensure that schools have the resources needed to support that learning. Documentation of the quality of classroom instruction is one indicator of opportunity to learn (see below).

Classroom Assessments

The changes in science education envisioned in the framework and the NGSS begin in the classroom. Instruction that reflects the goals of the framework and the NGSS will need to focus on developing students' skills and dispositions to use scientific and engineering practices to progress in their learning and to solve problems. Students will need to engage in activities that require the use of multiple scientific practices in developing a particular core idea and will need to experience the same practices in the context of multiple core ideas. The practices have to be used in concert with one another, for example, supporting an explanation with an argument or using mathematics to analyze data from an investigation.

Approaches to classroom assessment are discussed in detail in Chapter 4. Here, we emphasize their importance in an assessment system. As noted above, assessment systems have traditionally focused on large-scale external assessments, often to the exclusion of the role of classroom assessments. Achieving the goals of the framework and the NGSS will require an approach in which classroom assessment receives precedence. This change means focusing resources on the development and validation of high-quality materials to use as part of classroom teaching, learning, and assessment, complemented with a focus on developing the capacity of teachers to integrate assessments into instruction and to interpret the results to guide their teaching decisions.

In Chapter 4, we highlight examples of the types of classroom assessments that should be part of a system, and we emphasize that it is possible to develop assessment tasks that measure three-dimensional learning as envisioned in the

---

[5]Moderation is a set of processes designed to ensure that assessments are administered and scored in comparable ways. The aim of moderation is to ensure comparability; that is, that students who take the same subject in different schools or with different teachers and who attain the same standards will be recognized as being at the same level of achievement.

framework and the NGSS. It is worth noting, however, that each example is the product of multiple cycles of development and testing to refine the tasks, the scoring systems, and their interpretation and use by teachers. Thus, the development of high-quality classroom assessment that can be used for formative and summative purposes should be treated as a necessary and significant resource investment in classroom instructional supports, curriculum materials, and professional development for teachers.

Monitoring Assessments

In Chapter 5, we discuss assessments that are used to monitor or audit learning and note that it is not feasible to cover the full breadth and depth of the NGSS performance expectations for a given grade level with a single external (large-scale) assessment. The types of assessment tasks that are needed take time to administer, and several will be required in order to adequately sample the set of performance expectations for a given grade level. In addition, some practices, such as demonstrating proficiency in carrying out an investigation, will be difficult to assess in the conventional formats used for on-demand external assessments. Thus, states need to rely on a combination of two types of external assessment strategies for monitoring purposes: on-demand assessments (those developed outside the classroom and administered at a time mandated by the state) and classroom-embedded assessments (those developed outside the classroom and administered at a time determined by the district or school that fits the instructional sequence in the classroom).

A primary challenge in designing any monitoring assessment is in determining how to represent the domain to be assessed, given that (1) it will be difficult to cover all of the performance expectations for a given grade level without some type of sampling and (2) the monitoring assessments will be given to students who will have studied different curricula. There are various options: each has certain strengths but also some potential drawbacks.

One option is to sample the standards but not reveal which performance expectations will be covered by the assessment. This option encourages teachers to cover all of the material for a given grade, but it could lead to a focus on covering the full breadth of the material at the expense of depth. Another option is to make teachers and students aware of which subset of the performance expectations will be assessed in a particular time frame. Although this option encourages teachers to cover some performance expectations in depth, it also gives teachers an incentive to ignore areas that are not to be assessed. A third option is to make the sample choices public and to rotate the choices over time. This option helps to ensure that certain performance expectations are not consistently ignored, but it creates churning in instructional planning and also complicates possibilities for making comparisons across time.

It would also be possible to offer schools constrained choices from the full range of performance expectations, perhaps through attempts to prioritize the performance expectations. For example, schools might be encouraged to cover at least some particular number of disciplinary core ideas from given domains or offered a menu of sets of core ideas (perhaps with associated curriculum supports) from which to choose. Giving schools a constrained set of choices could allow for more flexibility, autonomy, and perhaps creativity. Providing them with a menu could also make it easier to ensure coherence across grade levels and to provide curriculum materials aimed at helping students meet key performance expectations.

Each option brings different advantages and disadvantages. Selecting the best option for a given state, district, or school context will depend on at least two other decisions. The first is whether to distribute the standards to be tested across the classroom-embedded component or in the on-demand component of the monitoring assessment: that is, which performance expectations would be covered in the classroom-embedded component and which in the on-demand component. The second is the extent to which there is state, district, or local school control over which performance expectations to cover. There is no strong *a priori* basis on which to recommend one option over the others, and thus states will need to use other decision criteria. We suggest two key questions that could guide a choice among possible strategies for representation of the standards: Will the monitoring assessment be used at the school, district, or state level? Which components of the monitoring assessment system (classroom embedded and on demand) will have choices associated with them?

Indicators of Opportunity to Learn

The work of identifying indicators of progress toward major goals for education in science, technology, engineering, and mathematics (STEM)—is already under-way and is described in a recent report *Monitoring Progress Toward Successful K-12 Education* (National Research Council, 2012b). The report describes a proposed set of indicators for K-12 STEM education that includes the goal of monitoring the extent to which state science assessments measure core concepts and practices and are in line with the new framework. The report includes a number of indicators that we think are key elements of a science accountability system: program inspections, student and teacher surveys, monitoring of teachers' professional development, and documentation of classroom assignments of students' work. These indicators would document such variables as time allocated to science teaching, adoption of instructional materials that reflect the NGSS and the framework's goals, and classroom coverage of content and practice outlined in these documents. Such indicators would be a critical tool for monitoring the equity of students' opportunities to learn.

A program of inspection of science classrooms could serve an auditing function, with a subset of schools sampled for an annual visit. The sample of schools could be randomly chosen, following a sampling design that accurately represents state-level science program characteristics. Schools with low scores on monitoring tests (or with low test scores relative to the performance expected based on other measures, such as achievement in other subject areas, socioeconomic status, etc.) would be more heavily sampled. Inspection would include documentation of resources (e.g., science space, textbooks, budgets for expendable materials), teacher qualifications, and time devoted to science instruction, including opportunities to engage in scientific and engineering practices. Peer review by highly qualified teachers (e.g., teachers with subject certification from the National Board for Professional Teaching Standards), who have had extensive training in the appropriate knowledge and skills for conducting such reviews, could be a component of an inspection program. These inspections would have to be designed not to be used in a punitive way, but to provide findings that could be used to guide schools' plans for improvement and support decisions about funding and resources.[6] We note that if such a program of inspection is implemented, forethought must be given to how recommendations for improvement can be supported.

Surveys of students and teachers could provide additional information about classrooms, as well as other variables such as students' level of engagement or teachers' content knowledge. The results of surveys used at selected grade levels together with data collected through a large-scale system component could also provide valuable background information and other data, and such surveys could be conducted online. Student surveys would have to be individually anonymous:

---

[6]Accreditation systems in the United States and other countries already use many of these strategies. For information about an organization that operates such a system in the United States and elsewhere, AdvancED, see http://www.advanc-ed.org/ [September 2013].

they would not include names but would be linked to schools. Student surveys could also be linked to teachers or to student demographic characteristics (e.g., race and ethnicity, language background, gender). If parallel versions of some questions are included on teacher and student questionnaires, then those responses could be compared. Questions could probe such issues as the amount of time spent on science instruction; opportunities for constructing explanations, argumentation, discussion, reasoning, model building, and formulation of alternative explanations; and levels of students' engagement and interest. Surveys for teachers could include questions about time spent in professional development or other professional learning opportunities.

The time provided for teacher collaboration and quality professional development designed to improve science teaching practices could also be monitored. Monitoring strategies could include teacher surveys completed at professional development events focused on science or school reporting of time and resources dedicated to supporting teachers' learning related to science.

Documentation of curriculum assignments or students' work might include portfolios of assignments and student work that could also provide information about the opportunity to learn (and might also be scored to provide direct information about student science achievement). The collected work could be rated for purposes of monitoring and improvement. Alternatively, the work could be used to provide an incentive for teachers to carefully consider aspects of the NGSS and the three-dimensional learning described in the framework (see Mitchell et al., 2004; Newmann et al., 1998; Newmann and Associates, 1996). Such a system of evaluation of the quality and demand of student assignments was used in Chicago and clearly showed that levels of achievement were closely tied to the intellectual demands of the work assigned to students (Newmann et al., 1998).

UNDERSTANDING THE SYSTEM COMPONENTS AND THEIR USES

As stated, a comprehensive science assessment system will include some measures that are closely linked to instruction and used primarily in classrooms for both formative and summative purposes (see Chapter 4). It will also include some measures designed to address specific monitoring purposes (see Table 5-1 in Chapter 5), including some that may be used as part of accountability policies. We recognize that adopting this approach would be a substantial change from what is currently done in most states and would require some careful consideration of how to assemble the components of an assessment system so that they provide useful and usable information for the wide variety of assessment purposes.

External on-demand assessments are more familiar to most people than other types of assessments. Moving from reliance on a single test to a comprehensive science assessment system to meet the NGSS goals is a big change. It will require policy makers to reconsider the role that assessment plays in the system: specifically, policy makers will need to consider which purposes require on-demand assessments that are given to all students in the state and which do not. We note that, for many purposes, there is no need to give the same test to all students in the state: matrix sampling, as discussed in Chapter 5, is a legitimate, viable, and often preferable option. And for other purposes, assessments that are more closely connected to classrooms and a specific curriculum are likely to be better choices than on-demand assessments.

Several connected sets of questions can guide thinking about the components of an assessment system:

- What is the purpose of the system and how will it serve to improve student learning?
    - For what purposes are assessment components needed?
    - How will the assessment and the use of the results help to improve student learning?
    - What results will be communicated to the various audiences?
    - How will the results be used, by whom, and what decisions will be based on them?
    - How will the results from different components relate to each other?
- What role will accountability play in the system?
    - Who will be held accountable for what?
    - How will accountability policies serve to improve student learning?
- Given the intended use of each of the assessment components in the system, at what levels (i.e., individual or group) will scores be needed?
    - Will information be needed about individuals or groups, such as those taught by particular teachers or who attend particular schools?
    - Do all students in the state need to take the same assessment component or can sampling of students and/or content be used?
- What level of standardization of different components is needed to support the intended use?
    - Do these uses require that certain assessment components be designed, administered, and scored by the state in a way that it is standardized across all school systems in the state?

o   Can school systems be given some choice about the exact nature of the assessment components, such as when they are given, and how they will be scored?

- What procedures will be used to monitor the quality of instruction and assessment in the system to ensure that students have access to high-quality instruction and the necessary resources?

The answers to these interrelated questions will help policy makers design an assessment system that meets their priorities.

EXAMPLES OF ALTERNATIVE SCIENCE ASSESSMENT SYSTEMS

In the following two sections we present a rough sketch of two alternative models for an assessment system.[7] As context for considering these alternative assessment models it is useful to note the ways in which they differ from the current system used in most states, the type of system that most students in this country experience. Currently, in most states, a single external (large-scale) assessment—designed or selected by the state—is given for monitoring purposes once in each grade span in elementary, middle, and high school. The assessment is composed predominantly of questions that assess factual recall. The assessment is given to all students and used to produce individual scores. Scores are aggregated to produce results at the group level. Classroom assessment receives relatively little attention in the current system, although this may vary considerably across schools depending on the resources available.

Although this is only a general sketch of the typical science assessment system in this country, it *is not* the type of system that we are recommending. In our judgment, this "default" system serves the purpose of producing numbers (test scores) that can be used to track science achievement on a limited range of content, but it cannot be used to assess learning in alignment with the vision of science learning in the framework or the NGSS.

As discussed above, the design of an assessment system should be based on a carefully devised plan that considers the purpose of each of the system components and how they will serve to improve student learning. The design should consider the types of evidence that are needed to achieve the intended purposes and

---

[7]These examples draw upon a presentation by Kathleen Scalise at the Invitational Research Symposium on Science Assessment sponsored by the Educational Testing Service, available at http://www.k12center.org/events/research-meetings/science.html [November 2013].

support the intended inference, and the types of assessment tasks needed to provide this evidence. In conceptualizing the system, we consider four critical aspects:

1. The system should include components designed to provide guidance for classroom teaching and learning.
2. It should include components designed for monitoring program effectiveness.
3. It should have multiple and convergent forms of evidence for use in holding schools accountable for meeting learning goals.
4. The various components should signify and exemplify important goals for student learning.

In the default system sketched above, results from large-scale standardized tests are used both for monitoring student learning and for program evaluation. The questions it includes signify the type of tasks students should be able to answer, which are not aligned with science learning envisioned in the framework and the NGSS. Test scores provide little information to guide instructional decision making. The examples in the next two sections provide only a rough sketch of two alternative systems—not all of the details that would need to be developed and worked out prior to implementation—but one can clearly see their differences with the current default model.

In Chapter 5, we describe two approaches to on-demand assessments (mixed-item formats with written responses and mixed-item formats with performance tasks) and three approaches to classroom-embedded assessment that could be used for monitoring purposes (replacement units, collections of performance tasks, and portfolios of work samples and work projects). In the system examples below, we explore ways to make use of these options in designing the monitoring assessment component of a system.

We assume that the assessment system would incorporate the advice offered in *Systems for State Science Assessment* (National Research Council, 2006) for designing a coherent system. That is, the system should be horizontally, vertically, and developmentally coherent. Horizontally, the curriculum, instruction, and assessment are aligned with the standards, target the same goals for learning, and work together to support students' developing science literacy (National Research Council, 2006, p. 5). Vertically, all levels of the education system—classroom, school, school district, and state—are based on a shared vision of the goals for science education, the purposes and uses of assessment, and what constitutes competent performance. Developmentally, the system takes account of how students' science understanding develops over time and the scientific content knowledge, abilities, and understanding that are needed for learning to progress at each stage of the process. (For further details about developing a comprehensive, coherent science assessment system, see National Research Council, 2006.)

We also assume that states and local education agencies would adopt NGSS-aligned curricula that incorporate the vision of science education conceptualized in the framework and would ensure that the system includes high-quality instructional materials and resources (including classroom assessments), that they would design suitable means of reporting the results of the assessments to appropriate audiences, and that teachers and administrators would receive comprehensive professional development so that they are well prepared for full implementation of a new system. Furthermore, we assume that available resources and professional development support the use of formative assessment as a regular part of instruction, relying on methods such as those described in Chapter 4. These features should be part of all science assessment systems. In the descriptions below, we focus on strategies for making use of the types of classroom and monitoring assessment strategies discussed in Chapters 4 and 5 of this report.

Example 1

In this model, the monitoring assessment would be given once in each grade span (elementary, middle, and high school, e.g., grades 4, 8, and 10) and would consist of two components. The first component would be one of the on-demand assessment options we suggest in Chapter 5. In this approach, a test that makes use of mixed-item formats including some constructed-response tasks (such as those currently used for the New England Common Assessment Program or on the New York state assessments or that were used in the past for the Maryland School Performance Assessment Program, see Chapter 5), would be used as an on-demand component. The second component would include several classroom-embedded assessments incorporated into replacement units (see Chapter 5).

For this model, the on-demand component would be administered in a way that makes use of both the fixed-form and matrix-sampling administration approaches. All students at a tested grade would take a common test form that uses selected-response and constructed-response questions (including some technology-enhanced questions, if feasible). Every student would also have to complete one of several performance assessment tasks, administered through a matrix-sampling design. The common, fixed-form test would yield score reports for individual students; the matrix-sampled portion would provide school-level scores.

Both parts of the monitoring assessment would be developed by the state. The state would determine when the on-demand assessment is given, but the district (or other local education agency) would make decisions about when the classroom-embedded assessment components would be scheduled and could select from among a set of options for the topics. Both parts of the monitoring assessment would be scored at the state level, although the state might decide to use teachers as scorers.

Although the assessments in the classroom-embedded component could be administered in a standardized way, one complication of this design is that it would be difficult to keep the assessments secure since they would be administered at different times of the school year. Thus, they would need to be designed in such a way that prior exposure to the assessment tasks would not interfere with measuring the intended constructs (performance expectations). In addition, further work would be needed on the best ways to combine results from the classroom-embedded component and the on-demand component.

Another decision would involve which performance expectations should be covered in the on-demand component and which ones would be covered in the classroom-embedded component. For example, the on-demand component could use currently available standardized tests for the disciplinary core ideas, adding in a set of complex tasks that also address a sampling of the scientific and engineering practices and crosscutting concepts. The classroom-embedded component could then assess a broader sample of the scientific and engineering practices and crosscutting concepts in the context of certain disciplinary core ideas.

In addition to the tasks used for the monitoring assessment, the state (or possibly a collaboration of states) would develop collections of tasks that could be used in the classroom to support formative and summative assessment purposes. The tasks would be designed to be aligned with the NGSS performance expectations and could be available for use in the classroom for a variety of purposes, such as to enliven instruction or to track progress (of course, the same tasks should not be simultaneously used for both). Teachers would be trained to score these tasks, and they would serve as examples for teachers to model as they develop their own assessments to use for classroom and instruction purposes.

Accountability policies would be designed to include indicators of opportunity to learn as discussed above, such as evidence that teachers have access to professional development and quality curricular materials and administrative supports, that they are implementing instruction and assessment in ways that align with the framework, and that all students have access to appropriate materials and resources.

Thus, in this example system, the classroom assessment component includes banks of tasks associated with specific performance expectations that demonstrate the learning goals for students and that are available for use in the classroom for instructional decision making. The monitoring component includes classroom-embedded and on-demand elements that allow for judgments about students' learning and for evaluation of program effectiveness. Results from the monitoring assessments, as well as indicators of opportunity to learn, would be used for holding districts and schools accountable for progress in meeting learning goals. The consistency of the information from the different parts of the assessment system would be used to monitor the system for variation in science learning outcomes across districts and schools.

Example 2

For this example, the on-demand component would consist of the mixed-item types option described in Chapter 5 that makes use of some selected-response questions and some short answer and extended constructed-response questions (such as the types of question formats on the advanced placement biology test discussed in Chapter 5 or some of the formats included in the taxonomy in Figure 5-6, in Chapter 5). The on-demand component would be administered as a fixed-form test that produces scores for individuals. Instead of replacement units, the classroom-embedded component would involve portfolios assembled to include examples of work in response to tasks specified by the state. The state would be in charge of scoring the assessments, including the portfolios, although it would be best if teachers were involved in the scoring.

This example shares some of the same complications as Example 1. Decisions will be needed as to which performance expectations will be covered in the on-demand assessment and which ones would be covered in the portfolios. It would also be difficult to maintain the security of the portfolio tasks if they are completed over the course of several weeks. In addition, assembling portfolios and evaluating the student work included in them is time and resource intensive. A research and development effort would be needed to investigate the best way to combine scores from the two types of assessments.

In addition to the monitoring assessment, portfolios could be used at each grade level to document students' progress. States or districts might collaborate to determine appropriate portfolio assignments and scoring rubrics; alternatively, an item bank of tasks and scoring rubrics could be developed to support classroom assessment. Decisions about the exact materials to be included in the portfolios would be determined by the state, the district, or the school. The portfolios would be scored at the district level by teachers who had completed training procedures as prescribed by the state for the monitoring assessment. The portfolios could be used as part of the data for assigning student grades.

As in Example 1, above, accountability would rely on results from the monitoring assessments as well as indicators of opportunity to learn. Samples of portfolios would be sent to the state for review of the quality of the assignments given to the students and the feedback teachers give them, providing one measure of opportunity to learn that could be combined with others, such as evidence that teachers have access to professional development and quality curricular materials and administrative supports, that they are implementing instruction and assessment in ways that align with the framework, and that all students have access to appropriate materials and resources.

Thus, in this system, the descriptions of materials to be included in portfolios exemplify the learning goals for students and are available to use in the classroom for instructional decision making. The external assessment allows for monitoring students' learning and evaluating program effectiveness. Results from the monitoring assessments as well as indicators of opportunity to learn would be used for holding schools accountable for meeting learning goals.

CONCLUSIONS AND RECOMMENDATIONS

In this chapter, we have discussed the importance of a systems approach to developing science assessments and described the system components that will be needed to adequately assess the breadth and depth of the NGSS.

CONCLUSION 6-1 A coherently designed multilevel assessment system is necessary to assess science learning as envisioned in the framework and the Next Generation Science Standards and provide useful and usable information to multiple audiences. An assessment system intended to serve accountability purposes and also support learning will need to include multiple components: (1) assessments designed for use in the classroom as part of day-to-day instruction, (2) assessments designed for monitoring purposes that include both on-demand and classroom-embedded components, and (3) a set of indicators designed to monitor the quality of instruction to ensure that students have the opportunity to learn science as envisioned in the framework. The design of the system and its individual components will depend on multiple decisions, such as choice of content and practices to be assessed, locus of control over administration and scoring decisions, specification of local assessment requirements, and the level and types of auditing and monitoring. These components and choices can lead to the design of multiple types of assessment systems.

We also note that designing reports of assessment results that are clear and understandable and useful for the intended purpose is an essential and critical aspect of the system design.

CONCLUSION 6-2 Assessment reporting is a critical element of a coherent system. How and to whom results will be reported are questions that need to be considered during the first stages of designing an assessment system because those answers will guide almost all subsequent decisions about the design of each of the system's assessment components and their relationship to each other.

Given the widespread concerns expressed above about adequate representation and coverage of the NGSS performance expectations, we make three recommendations related to the monitoring of student learning and the opportunity-to-learn functions that a state assessment system should be designed to support. Recommendations about the classroom assessment function are in Chapter 4; this function is one of the three pillars of any coherent state system even though it is not the primary focus of the recommendations in this chapter.

RECOMMENDATION 6-1 To adequately address the breadth and depth of the performance expectations contained in the Next Generation Science Standards, state and local policy makers should design their assessment systems so information used for monitoring purposes is obtained from both on-demand assessments developed by the state and a complementary set of classroom-embedded assessments developed either by the state or by districts, with state approval. To signify and make visible their importance, the monitoring assessment should include multiple performance-based tasks of three-dimensional science learning. When appropriate, computer-based technology should be used in monitoring assessments to broaden and deepen the range of performances demanded on the tasks in both the classroom-embedded and on-demand components.

The system design approach contained in Recommendation 6-1 will be necessary to fully cover the NGSS performance expectations for a given grade. Including a classroom-embedded component as part of the monitoring of student learning will demonstrate the importance of three-dimensional science learning and assessment to local educators while simultaneously providing them with examples and data to support ongoing improvements in instruction and learning.

RECOMMENDATION 6-2 States should routinely collect information to monitor the quality of classroom instruction in science, including the extent to which students have the opportunity to learn science in the ways called for in the framework, and the extent to which schools have the resources needed to support student learning. This information should be collected through inspections of school science programs, surveys of students and teachers, monitoring of teacher professional development programs, and documentation of curriculum assignments and student work.

For some monitoring purposes, individual student scores are not needed, only group-level scores. Whenever individual-level scores are not needed, the use of matrix-sampling procedures should be considered. Matrix sampling provides an efficient way to cover the domain more completely, can make it possible to use a wider array of performance-based tasks as well as equating techniques. In addition, hybrid models—that include some items or tasks common to all students and others that are distributed across students using matrix sampling—could also be used for monitoring functions (such as described above for Example 1).

RECOMMENDATION 6-3 In planning the monitoring elements of their system, state and local policy makers should design the on-demand and classroom-embedded assessment components so that they incorporate the use of matrix-sampling designs whenever appropriate (rather than requiring that every student take every item), especially for systems monitoring purposes. Variation in matrix-sampling designs—such as some that include a few questions or tasks common to all students and others that are distributed across students—should be considered for optimizing the monitoring process.

We caution against systems that place a primary focus on the monitoring assessment; rather, we encourage policy makers to take a balanced approach in allocating resources for each component of an assessment system. To ensure that all of the resources for developing assessments are not devoted to the monitoring component of the assessment system, we encourage policy makers to carefully consider the frequency with which the monitoring assessment is administered.

RECOMMENDATION 6-4 State and local policy makers should design the monitoring assessments in their systems so that they are administered at least once, but no more than twice, in each grade span (K-5, 6-8, 9-12), rather than in every grade every year.

Designing the links among the components of an assessment system, particularly between the on-demand components and the classroom-embedded assessment information, will be a key challenge in the development of an assessment system. Such links will be especially important if the information is to be used for accountability purposes. As noted throughout this report, if significant consequences are attached only to the on-demand assessments, instructional activities are likely to be focused on preparation for those assessments (teaching to the test). The kinds of learning objectives that can only be assessed using classroom-embedded assessments, such as student-designed investigations, are too important to exclude from the purview of the assessment monitoring and accountability system. Since the kinds of linkages that are needed have not yet been implemented in the United States, education decision makers face a challenge in trying to meet the goals of the Next Generation Science Standards.

RECOMMENDATION 6-5 Policy makers and funding agencies should support research on strategies for effectively using and integrating information from on-demand and classroom-embedded assessments for purposes of monitoring and accountability.

7

IMPLEMENTING A SCIENCE ASSESSMENT SYSTEM

The charge to this committee was to develop a plan for assessment that will reinforce and complement the dramatic changes to science education proposed in *A Framework for K-12 Science Education: Practices, Crosscutting Concepts, and Core Ideas* (National Research Council, 2012a, hereafter referred to as "the framework") and the *Next Generation Science Standards: For States, By States* (NGSS Lead States, 2013). We have emphasized throughout this report that both of these documents provide an opportunity to rethink the possibilities for using assessment to support learning. We recognize that changes of this order are extremely challenging, and our charge directed us specifically to discuss the feasibility and costs of our recommendations.

The guidance for developing a science assessment system discussed in Chapter 6 is based on the premise that states will need to tailor their plans to their own circumstances and needs. However, there are four major issues that will be important to implementation in any context. This chapter discusses these issues:

1. The development of a new assessment system will need to be undertaken gradually and phased in over time.
2. To be successful, a science assessment system will have to thoughtfully and consistently reflect the challenge of ensuring equity in the opportunity that students from diverse backgrounds have to demonstrate their knowledge and abilities. Meeting this challenge will require clear understanding of the opportunities all students have had to learn science and to be fairly assessed, in the new ways called for by the framework.

3. Technology will play a critical role in the implementation of any assessment system that is aligned with the framework and the Next Generation Science Standards (NGSS).

4. Every choice made in implementing a system will entail both costs and benefits and their tradeoffs, which will require careful analysis.

GRADUAL IMPLEMENTATION

In this report, we have presented examples of tasks that assess the three-dimensional science learning represented by the NGSS performance expectations, and examples of assessment strategies that can incorporate these tasks. We believe these examples will prove valuable to those who have the responsibility to plan and design new state science assessment systems, but they are only examples. Implementing new assessment systems will require substantial changes to current systems. Thus, state leaders and educators will need to be both patient and creative as they implement changes over time. They need to understand and plan for the development and implementation of new systems in stages, over a span of years.

A number of innovative assessment programs floundered in the 1990s in part because they were implemented far too rapidly (perhaps to meet political exigencies). In many cases, their developers were not given sufficient time to implement what were major changes or to make modifications as they learned from experience (McDonnell, 2004). Some veterans of these experiences have cited this as a key factor in the lack of sustainability of many such efforts (see National Research Council, 2010).

A new assessment system has to evolve alongside other elements that are changing. It will take time for the changes to curriculum, instruction, professional development, and the other components of science education envisioned in the framework and the NGSS to be developed and implemented. New modes of assessment will need to be coordinated with those other changes, both because what is needed has to be embedded in some way in curriculum and instruction and because there is little value in assessing students on material and kinds of learning that they have not had the opportunity to learn. Moreover, assessing knowledge through the application of practices is relatively new, particularly in the context of externally mandated assessments. States that adopt new science assessment systems will need time to further develop and test new types of tasks and technology and gather evidence of their efficacy and validity in measuring three-dimensional learning. These changes will also need to be accompanied by extensive changes in teacher professional development, at both the entry and continuing levels. Although these are all major changes, we note that many of them mirror those being proposed for assessment of English language arts and mathematics through the Race to the Top Assessment Program consortia.

As we emphasized in the discussion of our charge, striking the right balance with new assessments designed to measure rapidly changing curricula and instructional practices while also meeting a range of competing priorities will be challenging, and will require consideration of tradeoffs. Changes in curriculum, instruction, student performance expectations, and professional development will need to be carefully coordinated and then introduced and implemented in stages across grade levels. States will need to carefully plan and develop their own models for implementation. For example, some may want to begin at the kindergarten level and move upward by grade levels; others may choose another starting level, such as the beginning of middle school and move upwards (or downward) by grade levels. It is important to recognize that, in order to meet the performance expectations in the NGSS, students in higher grades will need to have had the necessary foundation in their earlier grades. States will need to expect and address these sorts of gaps, as they are currently doing with the Common Core State Standards in English language arts and mathematics.

It will be up to each state to determine the best way to gradually adapt their curricula. In many places, schools or districts have reduced the amount of science instruction offered in recent years, particularly in the early grades, in response to the accountability demands of the No Child Left Behind Act (NCLB) (see Center on Education Policy, 2007; Dorph et al., 2011; Griffith and Scharmann, 2008). Those jurisdictions will need to reintroduce science in the early grades—and review and revise the policies that have limited the time available for science—if they are to effectively implement the new standards. Frequently, schools that serve the most disadvantaged student populations are those in which the opportunity to learn science has been most reduced (Center on Education Policy, 2007; Dorph et al., 2011; Rennie Center for Education Research and Policy, 2008). Even in schools and districts that have maintained strong science programs at all grade levels, neither students nor teachers may have had experience with instruction that involves applying the practices as envisioned in the new framework and NGSS.

The cost of materials will also be a factor in the implementation of new approaches to science education, particularly at the elementary level. Many school districts in the United States use kit-based curriculum materials at the elementary levels, such as *Full Option Science Systems* (FOSS) and *Science and*

Technology for Children, which were developed in the early 1990s and aligned to AAAS benchmarks of the American Association for the Advancement of Science (1993, 2009) or to the *National Science Education Standards* (National Research Council, 1996). When combined with teacher training, these science kits have been valuable in the delivery of guided-inquiry instruction, but the materials will have to be revised and resequenced to align with the NGSS (Young and Lee, 2005). Developing the needed materials represents a significant investment for school districts.

Many states are already implementing the Common Core State Standards for English language arts and mathematics, which emphasize engaging students in classroom discourse across the disciplines. The new framework and the NGSS reflect the intention to integrate that approach with science learning: the integration will also take time and patience, especially in the many schools and districts in which there is little precedent on which to build.

Thus, states will need to both make some immediate changes and initiate a longer-term evolution of assessment strategies. Policy makers and educators will need to balance shorter- and longer-term assessment goals and to consider the effects of their goals and plans on each of the critical actors in teaching and assessment (e.g., the federal government, states, districts, schools, principals, teachers, parents, and students). Each component of the science education system—including instruction, curriculum and instructional materials, teacher education and professional development programs, assessment development, research, and education policy—will need to be adapted to an overall plan in a coordinated fashion. In terms of policy orientation, we emphasize again that a developmental path that is "bottom up" (i.e., grounded in the classroom), rather than "top down" (i.e., grounded in such external needs as accountability or teacher evaluation), is most likely to yield the evidence of student learning needed to support learning that is aligned with the framework's goals.

Although accountability is an important function of an assessment system, we believe that placing the initial focus on assessments that are as close as possible to the point of instruction will be the best way to identify successful strategies for teaching and assessing three-dimensional science learning. These strategies can then be the basis for the work of developing assessments at other levels, including external assessments that will be useful for purposes beyond the classroom. We recognize that we are calling on state and federal policy makers to change their thinking about accountability—to rethink questions about who should be held accountable for what and what kinds of evidence are most valuable for that task. States may have to temporarily forgo some accountability information if the new system is to have a chance to evolve as it needs to. Because this is a marked change, states that begin this approach will be breaking significant new ground, and there will be much to be learned from their experiences.

Continuing to use existing assessments will not support the changes desired in instruction, and thus interim solutions will be needed that can, simultaneously, satisfy federally mandated testing requirements and allow the space for change in classroom practice. Adapting new state assessment systems will require a lengthy transition period, just as the implementation of the NGSS in curriculum and instruction will require a gradual and strategic approach. A gradual approach will ease the transition process and strengthen the resulting system, both by allowing time for development and phasing in of curriculum materials aligned to the framework and by allowing all participants to gain familiarity and experience with new curricula and new kinds of instruction that address the three dimensions of the NGSS. Ideally, the transition period would be 5 years or more. We realize, however, that many states will face political pressures for much shorter timelines for implementation.

EQUITY AND FAIRNESS

A fundamental component of the framework's vision for science education is that all students can attain its learning goals. The framework and the NGSS both stress that this goal can only be reached if all students have the opportunity to learn in the new ways recommended in those documents. Achieving equity in the opportunity to learn science will be the responsibility of the entire system, but the assessment system can play a critical role by providing fair and accurate measures of the learning of all students. As we have noted, however, it will be challenging to strike the optimal balance in assessing students who are disadvantaged and students whose cultural and linguistic backgrounds may significantly influence their learning experiences in schools.

The K-12 student population in the United States is rapidly growing more diverse—culturally, linguistically, and in other ways (Frey, 2011). The 2010 U.S. census showed that while 36 percent of the total population are minorities, 45 percent of those who are younger than 19 are minorities (U.S. Census Bureau, 2012), and non-Asian minority students are significantly more likely to live in poverty than white or Asian students (Lee et al., 2013). The number of students who are considered limited English proficient doubled between 1993 and 2007, to 11 percent (Lee et al., 2013). Under any circumstances, assessing the learning of a very diverse student population requires attention to what those students have had the opportunity to learn and to the needs, perspectives, and modes of communication they bring to the classroom and to any assessment experience.

In the context of the recasting of science education called for by the framework and the NGSS, these issues of equity and fairness are particularly pressing. We argue in this report for a significantly broadened understanding of what assessment is and how it can be used to match an expanded conception of science learning. The framework and the NGSS stress the importance of such practices as analyzing and interpreting data, constructing explanations, and using evidence to defend an argument. Thus, the assessments we recommend present opportunities for students to engage in these practices. The implications for the equity of an assessment are complex, especially since there is still work to be done in devising the means of providing equitable opportunity to learn by participating in scientific practices that require significant discourse and writing.

Fairness is not a new concern in assessment. It can be described in terms of lack of bias in the assessment instrument, equitable treatment of test takers, and opportunity to learn tested material (American Educational Research Association, American Psychological Association, and National Council on Measurement in Education, 1999). It is important to note, however, that the presence of performance gaps among population groups does not necessarily signal that assessments are biased, unfair, or inequitable. Performance gaps on assessments may also signal important differences in achievement and learning among population groups, differences that will need to be addressed through improved teaching, instruction, and access to appropriate and adequate resources. A test that makes use of performance-based tasks may indeed reveal differences among groups that did not show up in tests that use other types of formats. NGSS-aligned assessments could be valuable tools for identifying those students who are not receiving NGSS-aligned instruction.

The changes to science education called for in the framework and the NGSS highlight the ways in which equity is integral to the definition of excellence. The framework stresses the importance of inclusive instructional strategies designed to engage students with diverse interests and backgrounds and points out that these principles should carry over into assessment design as well. It also notes that effective assessment must allow for the diverse ways in which students may express their developing understanding (National Research Council, 2012a, pp. 283, 290). The NGSS devotes an appendix to the discussion of "All Standards, All Students." It notes the importance of non-Western contributions to science and engineering and articulates three strategies for reaching diverse students in the classroom, which also apply to assessment (NGSS Lead States, 2013, Appendix D, p. 30):

1. Value and respect the experiences that all students bring from their backgrounds (e.g., homes and communities).
2. Articulate students' background knowledge (e.g., cultural or linguistic knowledge) with disciplinary knowledge.
3. Offer sufficient school resources to support student learning.

These principles offer a valuable addition to the well-established psychometric approaches to fairness in testing, such as statistical procedures to flag test questions that perform differently with different groups of students and may thus not measure all students' capability accurately (see e.g., American Educational Research Association, American Psychological Association, and National Council on Measurement in Education, 1999; Educational Testing Service, 2002; Joint Committee on Testing Practices, 2004). The principles are grounded in recent research that uses sociocultural perspectives to explore the relationships between individual learners and the environments in which they learn to identify some subtle but pervasive fairness issues (Moss et al., 2008). Although that research was primarily focused on different aspects of instruction and assessment, the authors have expanded the concept of opportunity to learn. In this view, opportunity to learn is a matter not only of what content has been taught and what resources were available, but also of (1) whether students' educational environments are sufficiently accessible and engaging that they can take advantage of the opportunities they have, (2) how they are taught, and (3) the degree to which the teacher was prepared to work with diverse student populations.

This research highlights the importance of respect for and responsiveness to diverse students' needs and perspectives. All students bring their own ways of thinking about the world when they come to school, based on their experiences, culture, and language (National Research Council, 2007). Their science learning will be most successful if curriculum, instruction, and assessments draw on and connect with these experiences and are accessible to students linguistically and culturally (Rosebery et al., 2010; Rosebery and Warren, 2008; Warren et al., 2001, 2005). It will not be easy for educators to keep this critical perspective in view while they are adapting to the significant changes called for by the framework and the NGSS. Moreover, given the current patterns of teacher experience and qualifications, it is likely that students in the most advantaged circumstances will be the first to experience science instruction that is guided by the framework and thus be prepared to succeed on new assessments. As states and districts begin to change their curricula and instruction and to adopt new assessments, they will need to pay careful attention to the ways in which students' experiences may vary by school and for different cultural groups. The information provided by new generations of assessments will only be meaningful to the extent that it reflects understanding of students' opportunities to learn in the new ways called for by the framework and educators find ways to elicit and make use of the diversity of students' interests and experiences. Monitoring of opportunity to learn, as we recommend (see Chapter 6), will thus be a critical aspect of any assessment system.

Because the language of science is specialized, language is a particular issue for the design of science assessments. To some extent, any content assessment will also be an assessment of the test takers' proficiency in the language used for testing (American Educational Research Association, American Psychological Association, and National Council on Measurement in Education, 1999). Both native English speakers and English-language learners who are unfamiliar with scientific terminology and various aspects of academic language may have difficulty demonstrating their knowledge of the material being tested if they have not also been taught to use these scientific modes of expression. Some researchers have suggested that performance tasks that involve hands-on activities are more accessible to students who are not proficient in English, but such tasks may still present complex linguistic challenges, and this issue should be considered in test design (Shaw et al., 2010).

We note that strategic use of technology may help to diminish these challenges. For example, technology can be used to provide flexible accommodations—such as translating, defining, or reading aloud words or phrases used in the assessment prompt or offering variable print size that allow students to more readily demonstrate their knowledge of the science being tested. One model for this approach is ONPAR (Obtaining Necessary Parity through Academic Rigor), a web resource for mathematics and science assessments that uses technology to minimize language and reading requirements and provide other modifications that make them accessible to all students.[1] However, more such examples are needed if the inclusive and comprehensive vision of the framework and the NGSS is to be realized.

---

[1]For details, see http://onpar.us/ [June 2013].

Researchers who study English-language learners also stress the importance of a number of strategies for engaging those students, and they note that these strategies can be beneficial for all students. For example, techniques used in literacy instruction can be used in the context of science learning. These strategies promote comprehension and help students build vocabulary so they can learn content at high levels while their language skills are developing (Lee, 2012; Lee et al., 2013).

Research illustrates ways in which attention to equity has been put into practice in developing assessments. One approach is known as universal test design, in which consideration of possible ways assessment format or structure might limit the performance of students is incorporated into every stage of assessment design and development (Thompson et al., 2002).[2] The concept of cultural validity has also been important. This idea takes the finding that "culture influences the ways in which people construct knowledge and create meaning from experience" (Solano-Flores and Nelson-Barber, 2001, p. 1) and applies it to both assessment design and development and to interpretation of assessment results (see also Basterra et al., 2011). Another approach is to provide specialized training for the people who will score the responses of culturally and linguistically diverse students to open-ended items (see Kopriva, 2008; Kopriva and Sexton, 1999).

Although building equity into assessment systems aligned with the framework and the NGSS poses challenges, it also presents opportunities. Equity in opportunity to learn is integral to the definition of excellence in those documents. Since significant research and development will be needed to support the implementation of the science assessment systems that are aligned with the framework and the NGSS, there is a significant opportunity for research and development on innovative assessment approaches and tasks that exemplify a view of excellence that is blended with the goals of equity. Much remains to be done: the new approaches called for in science education and in assessment should reflect the needs of an increasingly diverse student population. It will be important for those responsible for the design and development of science assessments to take appropriate steps to ensure that tasks are as accessible and fair to diverse student populations as possible. Individuals with expertise in the cultures, languages, and

---

[2]For more information, see Universally Designed Assessments from the National Center on Educational Outcomes, available at http://www.cehd.umn.edu/NCEO/TopicAreas/UnivDesign/UnivDesignTopic.htm [June 2013].

ethnicities of the student populations should be participants in assessment development and the interpretation and reporting of results.

We do not expect that any new approaches could, by themselves, eliminate inequity in science education. As we note earlier in this chapter, new assessments may very well reveal significant differences among groups of students, particularly because more advantaged schools and districts may implement the NGSS earlier and more effectively than less advantaged ones, at least in the early years. It will be important for test developers and researchers to fully explore any performance differences that become evident and to examine the factors that might contribute to them. For this type of research the appropriate types of data will have to be collected. This should include the material, human, and social resources available to support student learning, such as the indicators of opportunities to learn that we discuss in Chapter 6. Such studies might entail multivariate and hierarchical analyses of the assessment results so that factors influencing test scores can be better interpreted.[3]

TECHNOLOGY

Information and communications technology will be an essential component of a system for science assessment, as noted in the examples discussed throughout this report. Established and emerging technologies that facilitate the storage and sharing of information, audio and visual representation, and many other functions that are integral to the practice of science are already widely used in science instruction. As we have discussed, computer-based simulations allow students to engage in investigations that would otherwise be too costly, unsafe, or impractical. Simulations can also shorten the time needed to gather and display data (e.g., using computer-linked probes, removing repetitive steps through data spreadsheets and the application of algorithms) and give students access to externally generated datasets they can analyze and use as evidence in making arguments.

As we discuss in Chapter 5, technology enhances the options for designing assessment tasks that embody three-dimensional science learning. Technology can also support flexible accommodations that may allow English-language learners or students with disabilities to demonstrate their knowledge and skills. Students'

---

[3]These types of studies would not be attempts to do causal modeling, but a serious examination of sources of variance that might influences science scores especially when the scores are being used to make judgments about students and/or their teachers.

use of these options can be included as part of the data that are recorded and analyzed and used for future design purposes.[4]

Technology-based assessment in science is a fast-evolving area in which both the kinds of tasks that can be presented to students and the interface through which students interact with these tasks are changing. There are many interesting examples, but they do not yet comprise a fully evaluated set of strategies, so there are still questions to be answered about how technology-based tasks function. For example, tasks may ask students to manipulate variables in a simulation and interpret their observations or present data and data analysis tools for students to use in performing the given task. Students' familiarity and comfort with such simulations or tools will likely influence their ability to respond in the time allowed, regardless of their knowledge and skills. Therefore, it will be essential to ensure that students have experience with technology in the course of instruction, not just in the context of assessments. They need to gain familiarity with the interfaces and the requisite tools as part of their regular instruction before they are assessed using those tools, particularly when high stakes are attached to the assessment results. Moreover, the development of technology-based assessments needs to include extensive pilot testing so that students' reactions to the technology can be fully explored.[5]

COSTS

The charge to the committee included a discussion of the costs associated with our recommendations. Cost will clearly be an important constraint on implementing our recommendations and will influence the designs that states adopt. We strongly recommend that states adopt their new systems gradually and strategically, in phases, and doing so will be a key to managing costs. And as we discuss throughout the report, new and existing technologies offer possibilities for achieving assessment goals at costs lower than for other assessments, including performance

---

[4]We do not advocate that these data be used for the purpose of scaling the scores of students who make use of accommodations.

[5]One option for such pilot testing would be to develop an open-source database of simulations with a common interface style that can be used in both instruction and assessment, though this option would require a significant research and development effort. Another option would be to develop such resources as part of curriculum materials and give students the option of choosing assessment items that use the interface and simulation tools that match the curriculum that was used in their classrooms.

tasks. At the same time, much of what we recommend involves significant change and innovation, which will require substantial time, planning, and investment.

There is no simple way to generate estimates of what it might cost a state to transform its science assessment systems because each state will have a different starting point, a different combination of objectives and resources, and a different pace of change. The approach we recommend also means that assessments will be organically embedded in the science education system in a way that is fundamentally different from how assessments are currently understood and developed. An important advantage of the approach we recommend is that many assessment-related activities—such as task development and scoring moderation sessions in which teachers collaborate—will have benefits beyond their assessment function. Determining what portion of such an activity should be viewed as a new assessment cost, what portion replaces an older function, and what portion could fairly be treated as part of some other set of costs (e.g., professional development) may not be straightforward. It is possible to make some guesses, however, about ways in which the costs may be affected, and we see both significant potential savings and areas for which significant resources will be needed, particularly in the initial development phases.

Developing the design and implementation plan for the evolution to new assessment systems will require significant resources. The design and development of tasks of the kind we have described may be significantly more resource intensive than the design and development of traditional assessment tasks (such as tests composed of multiple-choice items), particularly in the early phases. And as we note above, research and experimentation will be needed over a period of years to complete the work of elaborating on the ideas reflected in the framework and the NGSS. There will also be ongoing costs associated with the administration and scoring of performance-based tasks.

A number of steps can be taken to help defray these costs. State collaboratives, such as the Race to the Top Assessment Program consortia for developing English language arts and mathematics assessments or the New England Common Assessment Program consortium for developing science assessments, can help to reduce development costs. Scoring costs may be reduced by using teachers as scorers (which also benefits their professional development) and by making use of automated scoring to the extent possible.[6] Integrating classroom-embedded assess-

---

[6]For a detailed analysis of costs associated with constructed-response and performance-based tasks, see Topol et al. (2010, 2013). Available: https://edpolicy.stanford.edu/sites/default/files/

ment into the system provides teacher-scored input, but the associated monitoring and moderating systems do have direct costs.

Looking at potential savings, the system design model we advocate will in many ways be more streamlined than the assessment programs most states are currently using. We recommend administering the monitoring assessments less frequently than is currently done in many states in many subjects (see Chapter 6). Much of what we recommend for classroom assessment will be integral to curriculum planning and professional development and thus both a shared cost and a shared resource with instruction. Furthermore, although the combination of classroom-based and monitoring assessments we propose may take longer to administer in the classroom, it will also be a benefit in terms of usefulness for instruction.[7]

We expect that costs will be most intense at the beginning of the process: as research and practice support increasing experience with the development of new kinds of tasks, the process will become easier and less costly. Each state, either on its own or in collaboration with other states, will have to build banks of tasks as well as institutional capacity and expertise.

Implementation of the NGSS will also bring states a number of advantages that have cost-saving implications. Because the NGSS will be implemented nationwide, states will be able to collaborate and to share resources, successful strategies, and professional development opportunities. This multistate approach is in stark contrast to the current approach, in which states have had distinct and separate science standards and have had to develop programs and systems to support science education in their states in relative isolation, often at significant cost and without the benefit of being able to build on successful models from other states.

The NGSS will also allow states to pilot professional development models in diverse and culturally varied environments, which could then be useful in other states or regions that have similar demographic characteristics.[8] The ways

---

publications/getting-higher-quality-assessments-evaluating-costs-benefits-and-investment-strategies.pdf [August 2013].

[7]It is a common mistake to see assessment as separate from the process of instruction rather than as an integral component of good instructional practice. Well-designed tasks and situations that probe students' three-dimensional science knowledge are opportunities for both student learning and student assessment. A substantial body of evidence shows that providing assessment opportunities in which students can reveal what they have learned and understood—to themselves, their peers, and their teachers—is far more beneficial to achievement than simply repeating the same content (Pashler et al., 2007 and Hinze et al., 2013).

[8]At least one such network to facilitate such interstate collaboration and mutual support is already operating. The Council of State Science Supervisors has organized meetings of BCSSE (Building Capacity for State Science Education) that included teams from more than 40 states in

in which states and school districts will be able to learn from one another and share successful models to support the systems of science education offer not only potentially substantial economies, but also an unparalleled opportunity to advance teaching and learning for all children.

RECOMMENDATIONS

Throughout the report we discuss and offer examples of practical ways to assess the deep and broad performance expectations outlined in the framework and the NGSS. However, we acknowledge the challenge of this new approach to assessment and building assessment systems. Implementing the recommended new approaches will require substantial changes, and it will take time. For the changes to be fully realized, all parts of the education system—including curriculum, instruction, assessment, and professional development—will need time to evolve. Thus, a key message is that each step needs to be taken with deliberation.

RECOMMENDATION 7-1 States should develop and implement new assessment systems gradually over time, beginning with what is both necessary and possible in the short term for instructional support and system monitoring while also establishing long-term goals to implement a fully integrated, technologically enhanced, coherent system of assessments.

RECOMMENDATION 7-2 Because externally developed assessments cannot, by design, assess the full range and breadth of the performance expectations in the Next Generation Science Standards (NGSS), they will have to focus on selected aspects of the NGSS (reflected as particular performance expectations or some other logical grouping structure). States should publicly reveal these assessment targets at least 1 year or more in advance of the assessment to allow teachers and students adequate opportunity to prepare.

As we discuss in Chapter 4, effective implementation of a new assessment system will require resources for professional development. Science instruction and

---

an ongoing collaboration about implementation issues for the NGSS and other new state standards for science, including but not limited to issues of assessment. Funding and resources to continue this networking will be an important investment to foster efficient learning from others in this multistate effort.

assessment cannot be successfully adapted to the new vision of science education without this element.

RECOMMENDATION 7-3 It is critically important that states include adequate time and material resources in their plans for professional development to properly prepare and guide teachers, curriculum and assessment developers, and others in adapting their work to the vision of the framework and the Next Generation Science Standards.

RECOMMENDATION 7-4 State and district leaders who commission assessment development should ensure that the plans address the changes called for by the framework and the Next Generation Science Standards. They should build into their commissions adequate provision for the substantial amounts of time, effort, and refinement that are needed to develop and implement such assessments, thus reflecting awareness that multiple cycles of design-based research will be necessary.

A fundamental component of the framework's vision for science education is that all students can attain its learning goals. The framework and the NGSS both stress that this goal can be reached only if all students have the opportunity to learn in the new ways recommended by those documents. Assessments will play a critical role in achieving this goal if they are designed to yield fair and accurate measures of the learning of all students. Careful attention to the diversity of the nation's student population will be essential in designing new science assessments.

RECOMMENDATION 7-5 Policy makers and other officials who are responsible for the design and development of science assessments should consider the multiple dimensions of diversity—including, but not limited to, culture, language, ethnicity, gender, and disability—so that the formats and presentation of tasks are as accessible and fair to diverse student populations as possible. Individuals with expertise in these areas should be integral participants in assessment development and in the interpretation and reporting of results.

As we discuss above, new assessments may reveal performance differences among groups for students, in part because more advantaged schools and districts might implement the NGSS earlier and more effectively than less advantaged ones. Data will need to be collected to support studies of any such performance differences.

RECOMMENDATION 7-6 Because assessment results cannot be fully understood in the absence of information about opportunities to learn what is tested, states should collect relevant indicators about opportunity to learn—including material, human, and social resources available to support student learning—to contextualize and validate the inferences drawn from the assessment results.

Information and communications technology will be an essential component of assessment systems designed to measure science learning as envisioned in the framework and the NGSS. Technology enhances options for designing assessment tasks that embody three-dimensional science learning, as well as strategies for making them more accessible to students with disabilities and English-language learners.

RECOMMENDATION 7-7 States should support the use of existing and emerging technologies in designing and implementing a science assessment system that meets the goals of the framework and the Next Generation Science Standards. New technologies hold particular promise for supporting the assessment of three-dimensional science learning, and for streamlining the processes of assessment administration, scoring, and reporting.

As the framework makes clear, assessment is a key element in the process of educational change and improvement. Done well, it can reliably measure what scientists, educators, and parents want students to know and be able to do, and it can help educators create the learning environments that support the attainment of those objectives. Done poorly, it will send the wrong message about what students know and can do, and it will skew the teaching and learning processes.

For K-12 science assessment, the framework and the NGSS provide an opportunity to rethink and redesign assessments so that they more closely align with a vision of science proficiency in which the practices of scientific reasoning are deeply connected with the understanding and application of disciplinary core ideas and crosscutting concepts. Defining in detail the nature of that understanding and developing valid ways to assess it present a substantial challenge for designing assessments. That challenge has begun to be met, as shown in the examples of such assessments, and there are tools, methods, and technologies now available to build on the work that has been done. If states, districts, researchers, and parents invest time and other resources in the effort, new science assessments that are well integrated with curriculum and instruction can be developed.

REFERENCES

Allen, R. (2012). *Developing the enabling context for school-based assessment in Queensland, Australia.* Washington, DC: The World Bank.

Almond, R.G., Steinberg, L.S., and Mislevy, R.J. (2002). A four-process architecture for assessment delivery, with connections to assessment design. *Journal of Technology, Learning, and Assessment, 1*(5). Available: http://www.bc.edu/research/intasc/jtla/journal/v1n5.shtml [June 2013].

Alonzo, A.C., and Gearhart, M. (2006). Considering learning progressions from a classroom assessment perspective. *Measurement: Interdisciplinary Research and Perspectives, 4*(1 and 2), 99-104.

American Association for the Advancement of Science. (2001). *Atlas of science literacy: Project 2061, Volume 1.* Washington, DC: Author.

American Association for the Advancement of Science. (2007). *Atlas of science literacy: Project 2061, Volume 2.* Washington, DC: Author.

American Association for the Advancement of Science. (2009, originally published 1993). *Benchmarks for Science Literacy.* New York: Oxford University Press.

American Educational Research Association, American Psychological Association, and National Council on Measurement in Education. (1999). *Standards for educational and psychological testing.* Washington DC: American Psychological Association.

Andrade, H., and Cizek, G.J. (Eds.). (2010). *Handbook of formative assessment.* New York: Taylor and Francis.

Association of Public and Land-grant Universities. (2011). *The common core state standards and teacher preparation: The role of higher education.* APPLU/SMTI, paper 2. Washington, DC: Author. Available: http://www.aplu.org/document.doc?id=3482 [May 2013].

Baker, E.L. (1994). Making performance assessment work: The road ahead. *Educational Leadership, 51*(6), 58-62.

Banilower, E.R., Fulp, S.L., and Warren, C.L. (2010). *Science: It's elementary. Year four evaluation report.* Chapel Hill, NC: Horizon Research.

Barton, K., and Schultz, G. (2012). *Using technology to assess hard-to-measure constructs in the CCSS and to expand accessibility: English language arts.* Paper presented at the Invitational Research Symposium on Technology Enhanced Assessments, Washington, DC. Available: http://www.k12center.org/events/research_meetings/tea.html [September 2013].

Basterra, M., Trumbul, E., and Solano-Flores, G. (2011). *Cultural validity in assessment: Addressing linguistic and cultural diversity.* New York: Routledge.

Baxter, G.P., and Glaser, R. (1998). The cognitive complexity of science performance assessments. *Educational Measurement: Issues and Practice, 17*(3), 37-45.

Bennett, R.E., and Bejar, I.I. (1998). Validity and automated scoring: It's not only the scoring. *Educational Measurement: Issues and Practice, 17*(4), 9-16.

Berland, L.K., and Reiser, B.J. (2009). Making sense of argumentation and explanation. *Science Education, 93*(1), 26-55.

Black, P., and Wiliam, D. (1998). Assessment and classroom learning. *Assessment in Education, 5*(1), 7-74.

Black, P., and Wiliam, D. (2010). Chapter 1: Formative assessment and assessment for learning. In J. Chappuis, *Seven strategies of assessment for learning.* New York: Pearson.

Black, P., Wilson, M., and Yao, S. (2011). Road maps for learning: A guide to the navigation of learning progressions. Measurement: *Interdisciplinary Research and Perspectives, 9*, 1-52.

Braun, H., Bejar, I.I., and Williamson, D.M. (2006). Rule-based methods for automated scoring: Applications in a licensing context. In D.M. Williamson, R.J. Mislevy, and I.I. Bejar (Eds.), *Automated scoring of complex tasks in computer-based testing* (pp. 83-122). Mahwah, NJ: Lawrence Erlbaum Associates.

Briggs, D., Alonzo, A., Schwab, C., and Wilson, M. (2006). Diagnostic assessment with ordered multiple-choice items. *Educational Assessment, 11*(1), 33-63.

Brown, N.J.S., Furtak, E.M., Timms, M.J., Nagashima, S.O., and Wilson, M. (2010) The evidence-based reasoning framework: Assessing scientific reasoning. *Educational Assessment, 15*(3-4), 123-141.

Buckley, B.C., and Quellmalz, E.S. (2013). Supporting and assessing complex biology learning with computer-based simulations and representations. In D. Treagust and C.-Y. Tsui (Eds.), *Multiple representations in biological education* (pp. 247-267). Dordrecht, Netherlands: Springer.

Buckley, J., Schneider, M., and Shang, Y. (2004). *The effects of school facility quality on teacher retention in urban school districts.* Washington, DC: National Clearinghouse for Educational Facilities. Available: http://www.ncef.org/pubs/teacherretention.pdf [December 2013].

Burke, R.J., and Mattis, M.C. (Eds.). (2007). *Woman and minorities in science technology, engineering, and mathematics: Upping the numbers.* Northampton, MA: Edward Elgar.

Bystydzienski, J.M., and Bird, S.R. (Eds.). (2006). *Removing barriers: Women in academic science, technology, engineering and mathematics.* Bloomington: Indiana University Press.

Camilli, G. (2006). Errors in variables: A demonstration of marginal maximum likelihood. *Journal of Educational and Behavioral Statistics, 31,* 311-325.

Camilli, G., and Shepard, L.A. (1994). *Methods for identifying biased test items.* Thousand Oaks, CA: Sage.

Catterall, J., Mehrens, W., Flores, R.G., and Rubin, P. (1998). *The Kentucky instructional results information system: A technical review.* Frankfort: Kentucky Legislative Research Commission.

Center on Education Policy. (2007). *Choices, changes, and challenges: Curriculum and instruction in the NCLB era.* Washington, DC: Author.

Claesgens, J., Scalise, K., Wilson, M., and Stacy, A. (2009). Mapping student understanding in chemistry: The perspectives of chemists. *Science Education, 93*(1), 56-85.

Clark- Ibañez, M. (2004). Framing the social world through photo-elicitation interviews. *American Behavioral Scientist, 47*(12), 1507-1527.

College Board. (2009). *Science: College Board standards for college success.* New York: Author.

College Board. (2011). *AP biology: Curriculum framework 2012-2013.* New York: Author.

College Board. (2012). *AP biology course and exam description effective fall 2012.* New York: Author

College Board. (2013a). *AP biology 2013 free-response questions.* New York: Author. Available: https://secure-media.collegeboard.org/ap-student/pdf/biology/ap-2013-biology-free-response-questions.pdf [December 2013].

College Board. (2013b). *AP biology 2013 scoring guidelines.* New York: Author. Available: http://media.collegeboard.com/digitalServices/pdf/ap/apcentral/ap13_biology_q2.pdf [December 2013].

Corcoran, T., Mosher, F.A., and Rogat, A. (2009). *Learning progressions in science.* Philidelphia, PA: Consortium for Policy Research in Education.

Darling-Hammond, L., Herman, J., Pellegrino, J., Abedi, J., Aber, J.L., Baker, E., Bennett, R., Gordon, E., Haertel, E., Hakuta, K., Ho, A., Linn, R.L., Pearson, P.D., Popham, J., Resnick, L., Schoenfeld, A.H., Shavelson, R., Shepard, L.A., Shulman, L., and Steele, C.M. (2013). *Criteria for high-quality assessment.* Stanford, CA: SCOPE, CRESST, and Learning Sciences Research Institute.

Davis, E.A., Petish, D., and Smithey, J. (2006). Challenges new science teachers face. *Review of Educational Research, 76*(4), 607-651.

Dee, T.S., Jacob, B.A., and Schwartz, N.L. (2013). The effects of NCLB on school resources and practices. *Educational Evaluation and Policy Analysis, 35*(2), 252-279.

Deng, N., and Yoo, H. (2009). *Resources for reporting test scores: A bibliography for the assessment community.* Prepared for the National Council on Measurement in Education, Center for Educational Measurement, University of Massachusetts, Amherst. Available: http://ncme.org/linkservid/98ADCCAD-1320-5CAE-6ED4F61594850156/showMeta/0/ [April 2014].

Dietel, R. (1993). What works in performance assessment? In Proceedings of the 1992 CRESST Conference. *Evaluation Comment Newsletter* [Online]. Available: http://www.cse.ucla.edu/products/evaluation/cresst_ec1993_2.pdf [October 2013].

diSessa, A.A. (2004). Metarepresentation: Native competence and targets for instruction. *Cognition and Instruction, 22*, 293-331.

diSessa, A.A., and Minstrell, J. (1998). Cultivating conceptual change with benchmark lessons. In J.G. Grceno and S. Goldman (Eds.), *Thinking practices.* Mahwah, NJ: Lawrence Erlbaum Associates.

Dorph, R., Shields, P., Tiffany-Morales, J., Hartry, A., and McCaffrey, T. (2011). *High hopes— few opportunities: The status of elementary science education in California.* Sacramento, CA: The Center for the Future of Teaching and Learning at WestEd.

Draney, K., and Wilson, M. (2008). A LLTM approach to the examination of teachers' ratings of classroom assessment tasks. *Psychology Science, 50*, 417.

Dunbar, S., Koretz, D., and Hoover, H.D. (1991). Quality control in the development and use of performance assessment. *Applied Measurement in Education, 4*(4), 289-303.

Duschl, R., and Gitomer, D. (1997). Strategies and challenges to changing the focus of assessment and instruction in science classrooms. *Educational Assessment, 4*, 37-73.

Educational Testing Service. (2002). *ETS standards for quality and fairness.* Providence, NJ: Author. Available: http://www.ets.org/s/about/pdf/standards.pdf [October 2013].

Ericsson, K.A., and Simon, H.A. (1984). *Protocol analysis: Verbal reports as data.* Cambridge, MA: Bradford Books/MIT Press.

Ferrara, S. (2009). *The Maryland School Performance Assessment Program (MSPAP) 1991-2002: Political considerations.* Paper prepared for the Workshop of the Committee on Best Practices in State Assessment Systems: Improving Assessment while Revisiting Standards, December 10-11, National Research Council, Washington, DC. Available: http://www7.nationalacademies.org/bota/Steve%20Ferrara.pdf [September 2010].

Frey, W.H. (2011). *America's diverse future: Initial glimpses at the U.S. child population from the 2010 Census.* Brookings Series: State of Metropolitan America Number 26 of 62, April. Available: http://www.brookings.edu/~/media/research/files/papers/2011/4/06%20census%20diversity%20frey/0406_census_diversity_frey.pdf [October 2013].

Gobert, J.D. (2000). A typology of causal models for plate tectonics: Inferential power and barriers to understanding. *International Journal of Science Education, 22*(9), 937-977.

Gobert, J.D. (2005). The effects of different learning tasks on model-building in plate tectonics: Diagramming versus explaining. *Journal of Geoscience Education, 53*(4), 444-455.

Gobert, J.D., and Clement, J.J. (1999). Effects of student-generated diagrams versus student-generated summaries of conceptual understanding of causal and dynamic knowledge in plate tectonics. *Journal of Research in Science Teaching, 36*, 36-53.

Gobert, J.D., and Pallant, A. (2004). Fostering students' epistemologies of models via authentic model-based tasks. *Journal of Science Education and Technology, 13*(1), 7-22.

Gobert, J.D., Horwitz, P., Tinker, B., Buckley, B., Wilensky, U., Levy, S., and Dede, C. (2003). *Modeling across the curriculum: Scaling up modeling using technology.* Proceedings of the Twenty-Fifth Annual Meeting of the Cognitive Science Society. Available: http://ccl.sesp.northwestern.edu/papers/2003/281.pdf [December 2013].

Gong, B., and DePascale, C. (2013). *Different but the same: Assessment "comparability" in the era of the common core state standards.* White paper prepared for the Council of Chief State School Officers, Washington, DC.

Goodman, D.P., and Hambleton, R.K. (2004). Student test score reports and interpretive guides: Review of current practices and suggestions for future research. *Applied Measurement in Education, 17*(2), 145-220.

Gotwals, A.W., and Songer, N.B. (2013). Validity evidence for learning progession-based assessment items that fuse core disciplinary ideas and science practices. *Journal of Research in Science Teaching, 50*(5), 597-626.

Griffith, G., and Scharmann, L. (2008). Initial impacts of No Child Left Behind on elementary science education. *Journal of Elementary Science Education, 20*(3), 35-48.

Haertel, E. (2013). How is testing supposed to improve schooling? *Measurement: Interdisciplinary Research and Perspectives, 11*(1-2), 1-18.

Haertel, E., Beauregard, R., Confrey, J., Gomez, L., Gong, B., Ho, A., Horwitz, P., Junker, B., Pea, R, and Shepard, L. (2012). *NAEP: Looking ahead—leading assessment into the future. Recommendations to the Commissioner.* Washington, DC: National Center for Education Statistics.

Hambleton, R.K., and Slater, S. (1997). Reliability of credentialing examinations and the impact of scoring models and standard-setting policies. *Applied Measurement in Education, 13*, 19-38.

Hambleton, R.K., Impara J., Mehrens W., and Plake B.S. (2000). *Psychometric review of the Maryland School Performance Assessment Program (MSPAP)*. Baltimore, MD: Abell Foundation.

Hambleton, R.K., Jaeger, R.M., Koretz, D., Linn, R.L., Millman, J., and Phillips, S.E. (1995). *Review of the measurement quality of the Kentucky instructional results information system, 1991-1994*. Report prepared for the Office of Educational Accountability, Kentucky General Assembly.

Hamilton, L.S., Nussbaum, E.M., and Snow, R.E. (1997). Interview procedures for validating science assessments. *Applied Measurement in Education, 10*(2), 191-200.

Hamilton, L.S., Stecher, B.M., and Klein, S.P. (Eds.) (2002). *Making sense of test-based accountability in education*. Santa Monica, CA: RAND.

Hamilton, L.S., Stecher, B.M., and Yuan, K. (2009). *Standards-based reform in the United States: History, research, and future directions*. Washington, DC: Center on Education Policy.

Harmon, M., Smith, T.A., Martin, M.O., Kelly, D.L., Beaton, A.E., Mullis, I.V.S., Gonzalez, E.J. and Orpwood, G. (1997). *Performance assessment in IEA's third international mathematics and science study*. Chestnut Hill, MA: Center for the Study of Testing, Evaluation, and Educational Policy, Boston College.

Heritage, M. (2010). *Formative assessment and next-generation assessment systems: are we losing an opportunity?* Los Angeles, CA: Council of Chief State School Officers.

Hill, R.K., and DePascale, C.A. (2003). Reliability of no child left behind accountability designs. *Educational Measurement: Issues and Practices, 22*(3), 12-20.

Hinze, S.R., Wiley, J., and Pellegrino, J.W. (2013). The importance of constructive comprehension processes in learning from tests. *Journal of Memory and Language, 69(2)*, 151-164.

Ho, A.D. (2013). The epidemiology of modern test score use: Anticipating aggregation, adjustment, and equating. *Measurement: Interdisciplinary Research and Perspectives, 11*, 64-67.

Holland, P.W., and Dorans, N.J. (2006). Linking and equating. In R.L. Brennan (Ed.), *Educational measurement* (4th ed., pp. 187-220). Westport, CT: Praeger.

Holland, P.W., and Wainer, H. (Eds.). (1993). *Differential item functioning*. Hillsdale, NJ: Lawrence Erlbaum Associates.

Hoskens, M., and Wilson, M. (2001). Real-time feedback on rater drift in constructed response items: An example from the Golden State Examination. *Journal of Educational Measurement, 38*, 121-145.

Huff, K., Steinberg, L., and Matts, T. (2010). The promises and challenges of implementing evidence-centered design in large-scale assessment. *Applied Measurement in Education, 23*(4), 310-324.

Huff, K., Alves, C., Pellegrino, J., and Kaliski P. (2012). Using evidence-centered design task models in automatic item generation. In M. Gierl and T. Haladyna (Eds.), *Automatic item generation.* New York: Informa UK.

Intergovernmental Panel on Climate Change (2007). *Climate change 2007: Synthesis report.* Geneva, Switzerland: Author.

International Association for the Evaluation of Educational Achievement. (2013). *International computer and information literacy study: Assessment framework.* Amsterdam, the Netherlands: Author.

International Baccalaureate Organization. (2007). *Diploma programme: Biology guide.* Wales, UK: Author.

International Baccalaureate Organization. (2013). *Handbook of procedures for the diploma programme, 2013.* Wales, UK: Author.

Jaeger, R.M. (1996). *Reporting large-scale assessment results for public consumption: Some propositions and palliatives.* Presented at the 1996 Annual Meeting of the National Council on Measurement in Education, April, New York.

Joint Committee on Testing Practices (2004). *Code of fair testing practices in education.* Washington DC: American Psychological Association. Available: http://www.apa.org/science/programs/testing/fair-code.aspx [December 2013].

K-12 Center at Educational Testing Service. (2013). *Now underway: A step change in K-12 Testing.* Princeton, NJ: Educational Testing Service. Available: http://www.k12center.org/rsc/pdf/a_step_change_in_k12_testing_august2013.pdf [December 2013].

Kane, M.T. (2006). Validation. In R. L. Brennan (Ed.), *Educational measurement: Fourth edition* (pp. 17-64). Westport, CT: Praeger.

Kane, M.T. (2013). Validation as a pragmatic, science activity. *Journal of Educational Measurement, 50*(1), 115-122.

Kennedy, C.A. (2012a). *PBIS student assessment on earth systems concepts: Test and item analysis.* Berkeley, CA: KAC Consulting.

Kennedy, C.A. (2012b). *PBIS student assessment on energy concepts: Test and item analysis.* Unpublished Report. Berkeley, CA: KAC Consulting.

Kingston, N., and Nash, B. (2011). Formative assessment: A meta-analysis and a call for research. *Educational Measurement: Issues and Practice, 30*(4), 28-37.

Klein, S.P., McCaffrey, D., Stecher, B., and Koretz, D. (1995). The reliability of mathematics portfolio scores: Lessons from the Vermont experience. *Applied Measurement in Education, 8*(3), 243-260.

Kolen, M.J., and Brennan, R.L. (2004). *Test equating, linking, and scaling: Methods and practices* (2nd ed.). New York: Springer-Verlag.

Kopriva, R., and Sexton, U.M. (1999). *Guide to scoring LEP student responses to open-ended science items.* Washington, DC: Council of Chief State School Officers.

Kopriva, R.J. (2008). *Improving testing for English language learners.* New York: Routledge.

Koretz, D. (2005). *Alignment, high stakes, and inflation of test scores.* CSE Report 655. Los Angeles: National Center for Research on Evaluation, Standards, and Student Testing, Center for the Study of Evaluation, Graduate School of Education & Information Studies, University of California, Los Angeles.

Koretz, D. (2008). *Measuring up: What educational testing really tells us.* Cambridge, MA: Harvard University Press.

Koretz, D., McCaffrey, D., Klein, S., Bell, R., and Stecher, B. (1992a). *The reliability of scores from the 1992 Vermont portfolio assessment program: Interim report.* CSE Technical Report 355. Santa Monica, CA: RAND Institute on Education and Training.

Koretz, D., Stecher, B., and Deibert, E. (1992b). *The Vermont portfolio assessment program: Interim report on implementation and impact, 1991-1992 school year.* CSE Technical Report 350. Santa Monica, CA: RAND and Los Angeles: Center for Research on Evaluation, Standards and Student Testing, University of California.

Koretz, D., Klein, S., McCaffrey, D., and Stecher, B. (1993a). *Interim report: The reliability of Vermont portfolio scores in the 1992-1993 school year.* CSE Technical Report 370. Santa Monica, CA: RAND and Los Angeles: Center for Research on Evaluation, Standards and Student Testing, University of California.

Koretz, D., Stecher, B., Klein, S., McCaffrey, D., and Deibert, E. (1993b). *Can portfolios assess student performance and influence instruction? The 1991-1992 Vermont experience.* CSE Technical Report 371. Santa Monica, CA: RAND and Los Angeles: Center for Research on Evaluation, Standards and Student Testing, University of California.

Koretz, D., Stecher, B., Klein, S., and McCaffrey, D. (1994). *The evolution of a portfolio program: The Impact and quality of the Vermont program in its second year (1992-93),* CSE Technical Report 385. Los Angeles: Center for Research on Evaluation, Standards and Student Testing, University of California.

Krajcik, J., and Merritt, J. (2012). Engaging students in scientific practices: What does constructing and revising models look like in the science classroom? Understanding a framework for K-12 science education. *The Science Teacher, 79,* 38-41.

Krajcik, J., Slotta, J., McNeill, K.L., and Reiser, B (2008a). Designing learning environments to support students constructing coherent understandings. In Y. Kali, M.C. Linn, and J.E. Roseman (Eds.), *Designing coherent science education: Implications for curriculum, instruction, and policy.* New York: Teachers College Press.

Krajcik, J., McNeill, K.L. and Reiser, B. (2008b). Learning-goals-driven design model: Curriculum materials that align with national standards and incorporate project-based pedagogy. *Science Education, 92*(1), 1-32.

Krajcik, J., Reiser, B.J., Sutherland, L.M., and Fortus, D. (2013). *Investigating and questioning our world through science and technology.* Second ed. Greenwich, CT: Sangari Active Science.

Labudde, P., Nidegger, C., Adamina, M. and Gingins, F. (2012). The development, validation, and implementation of standards in science education: Chances and difficulties in the Swiss project HarmoS. In S. Bernholt, K. Neumann, and P. Nentwig (Eds), *Making It Tangible: Learning Outcomes in Science Education* (pp. 237-239). Munster, Germany: Waxmann.

Lee, O. (2012). *Next generation science standards for English language learners.* Presentation prepared for the Washington Association of Bilingual Education, May 11, New York University.

Lee, O., Quinn, H., and Valdés, G. (2013). Science and language for English language learners in relation to next generation science standards and with implications for common core state standards for English language arts and mathematics. *Educational Researcher, 42*(4), 223-233.

Lehrer, R. (2011). *Learning to reason about variability and chance by inventing measures and models.* Paper presented at the National Association for Research in Science Teaching, Orlando, FL.

Lehrer, R.L., and Schauble, L. (2012). Seeding evolutionary thinking by engaging children in modeling its foundations. *Science Education, 96*(4), 701-724.

Lehrer, R.L., Kim, M., and Schauble, L. (2007). Supporting the development of conceptions of statistics by engaging students in modeling and measuring variability. *International Journal of Computers for Mathematics Learning, 12*, 195-216.

Lehrer, R., Kim, M.J., and Jones, S. (2011). Developing conceptions of statistics by designing measures of distribution. *International Journal on Mathematics Education (ZDM), 43*(5), 723-736.

Lehrer, R., Kim, M-J., Ayers, E., and Wilson, M. (2013). Toward establishing a learning progression to support the development of statistical reasoning. In J. Confrey and A. Maloney (Eds.), *Learning over Time: Learning trajectories in mathematics education.* Charlotte, NC: Information Age.

Linn, R.L., Baker, E.L., and Dunbar, S.B. (1991). Complex performance-based assessment: Expectations and validation criteria. *Educational Researcher, 20*(8), 15-21.

Linn, R.L., Burton, E.L., DeStafano, L., and Hanson, M. (1996). Generalizability of new standards project 1993 pilot study tasks in mathematics. *Applied Measurement in Education, 9*(3), 201-214.

Masters, G.N., and McBryde, B., (1994) *An investigation of the comparability of teachers' assessment of student folios.* Research report number 6. Brisbane, Australia: Queensland Tertiary Entrance Procedures Authority.

Michaels, S., and O'Connor, C. (2011). *Problematizing dialogic and authoritative discourse, their coding in classroom transcripts and realization in the classroom.* Paper presented at ISCAR, the International Society for Cultural and Activity Research. Rome, Italy. September 7.

Marion, S., and Shepard, L. (2010). *Let's not forget about opportunity to learn: Curricular support for innovative assessments.* Dover, NH: The National Center for the Improvement of Educational Assessment, Center for Assessment. Available: http://www.nciea.org/publication_PDFs/Marion%20%20Shepard_Curricular%20units_042610.pdf [June 2013].

McDonnell, L.M. (2004). *Politics, persuasion, and educational testing.* Cambridge, MA: Harvard University Press.

McNeill, K.L. and Krajcik, J. (2008). Scientific explanations: Characterizing and evaluating the effects of teachers' instructional practices on student learning. *Journal of Research in Science Teaching, 45*(1), 53-78.

Messick, S. (1989). Validity. In R.L. Linn (Ed.), *Educational Measurement* (3rd ed., pp. 13- 104). New York: Macmillan.

Messick, S. (1993). *Foundations of validity: Meaning and consequences in psychological assessment.* Princeton, NJ: Educational Testing Service.

Messick, S. (1994). The interplay of evidence and consequences in the validation of performance assessments. *Education Researcher, 23*(2), 13-23.

Minstrell, J., and Kraus, P. (2005). Guided inquiry in the science classroom. In M.S. Donovan and J.D. Bransford (Eds.), *How Students Learn: History, Mathematics, and Science in the Classroom.* Washington, DC: The National Academies Press.

Minstrell, J., and van Zee, E. H. (2003). Using questioning to assess and foster student thinking. In J. M. Atkin and J. E. Coffey (Eds.), *Everyday assessment in the science classroom.* Arlington, VA: National Science Teachers Association.

Mislevy, R.J. (2007). Validity by design. *Educational Researcher, 36*, 463-469.

Mislevy, R.J., and Haertel, G. (2006). Implications for evidence centered design for educational assessment, *Educational Measurement: Issues and Practice, 25*, 6-20.

Mislevy, R.J., Almond, R.G., and Lukas, J.F. (2003). *A brief introduction to evidence-centered design.* Princeton, NJ: Educational Testing Service.

Mislevy, R.J., Steinberg, L.S., and Almond, R.A. (2002). Design and analysis in task-based language assessment. *Language Testing, 19*, 477-496.

Mitchell, K.J., Murphy, R.F., Jolliffe, D., Leinwand, S., and Hafter, A. (2004). *Teacher assignments and student work as measures of opportunity to learn.* Menlo Park, CA: SRI International.

Moss, P.A., Pullin, D.C., Gee, J.P., Haertel, E.H., and Young, L.J. (Eds.). (2008). *Assessment, equity, and opportunity to learn.* Cambridge, UK: Cambridge University Press.

National Academy of Engineering and National Research Council. (2009). *Engineering in K-12 education.* Committee on K-12 Engineering Education, L. Katehi, G. Pearson, and M. Feder (Eds.). Washington, DC: The National Academies Press.

National Assessment Governing Board. (2009). *Science framework for the 2009 national assessment of educational progress.* Washington, DC: U.S. Department of Education. Available: http://www.nagb.org/content/nagb/assets/documents/publications/frameworks/science-09.pdf [May 2013].

National Research Council. (1996). *National science education standards.* National Committee for Science Education Standards and Assessment. National Committee on Science Education Standards and Assessment, Board on Science Education, Division of Behavioral and Social Sciences and Education, National Research Council. Washington, DC: National Academy Press.

National Research Council. (2000). *How people learn: Brain, mind, experience, and school.* Committee on Developments in the Science of Learning. J.D. Bransford, A.L. Brown, and R.R. Cocking (Eds.). Committee on Learning Research and Educational Practice, M.S. Donovan, J.D. Bransford, and J.W. Pellegrino (Eds.). Commission on Behavioral and Social Sciences and Education. Washington, DC: National Academy Press.

National Research Council. (2001). *Knowing what students know: The science and design of education assessment.* Committee on the Foundations of Assessment. J.W. Pellegrino, N. Chudowsky, and R. Glaser (Eds.). Board on Testing and Assessment, Center for Education, Division of Behavioral and Social Sciences and Education. Washington, DC: The National Academies Press.

National Research Council. (2003). *Assessment in support of instruction and learning: Bridging the gap between large-scale and classroom assessment: Workshop report.* Committee on Assessment in Support of Instruction and Learning, Committee on Science Education K-12. Board on Testing and Assessment, Committee on Science Education K-12, Mathematical Sciences Education Board, Center for Education. Division of Behavioral and Social Sciences and Education. Washington, DC: The National Academies Press.

National Research Council. (2005). *America's lab report: Investigations in high school science.* Committee on High School Science Laboratories: Role and Vision, S.R. Singer, M.L. Hilton, and H.A. Schweingruber (Eds.). Board on Science Education, Center for Education. Division of Behavioral and Social Sciences and Education. Washington, DC: The National Academies Press.

National Research Council. (2006). *Systems for state science assessment.* Committee on Test Design for K-12 Science Achievement. M.R. Wilson and M.W. Bertenthal (Eds.). Board on Testing and Assessment, Center for Education, Division of Behavioral and Social Sciences and Education. Washington, DC: The National Academies Press.

National Research Council. (2007). *Taking science to school: Learning and teaching science in grades K-8.* Committee on Science Learning, Kindergarten Through Eighth Grade. R.A. Duschl, H.A. Schweingruber, and A.W. Shouse (Eds.). Board on Science Education, Center for Education. Division of Behavioral and Social Sciences and Education. Washington, DC: The National Academies Press.

National Research Council. (2009). *Learning science in informal environments: People, places, and pursuits.* Committee on Learning Science in Informal Environments. P. Bell, B. Lewenstein, A.W. Shouse, and M.A. Feder (Eds.). Board on Science Education, Center for Education. Division of Behavioral and Social Sciences and Education. Washington, DC: The National Academies Press.

National Research Council. (2010). *State assessment systems: Exploring best practices and innovations: Summary of two workshops.* A. Beatty, Rapporteur, Committee on Best Practices for State Assessment Systems: Improving Assessment While Revisiting Standards. Board on Testing and Assessment. Division of Behavioral and Social Sciences and Education. Washington, DC: The National Academies Press.

National Research Council. (2011a). *Expanding underrepresented minority participation: America's science and technology talent at the crossroads.* Committee on Underrepresented Groups and the Expansion of the Science and Engineering Workforce Pipeline. Washington, DC: The National Academies Press.

National Research Council. (2011b). *Incentives and test-based accountability in education.* M. Hout and S.W. Elliott (Eds.). Committee on Incentives and Test-Based Accountability in Public Education. Board on Testing and Assessment, Division of Behavioral and Social Sciences and Education. Washington, DC: The National Academies Press.

National Research Council. (2012a). *A framework for K-12 science education: Practices, crosscutting concepts, and core ideas.* Committee on Conceptual Framework for the New K-12 Science Education Standards. Board on Science Education. Division of Behavioral and Social Sciences and Education. Washington, DC: The National Academies Press.

National Research Council. (2012b). *Monitoring progress toward successful K-12 STEM education: A nation advancing?* Committee on the Evaluation Framework for Successful K-12 STEM Education. Board on Science Education and Board on Testing and Assessment, Division of Behavioral and Social Sciences and Education. Washington, DC: The National Academies Press.

National Task Force on Teacher Education in Physics. (2010). *National task force on teacher education in physics: Report synopsis.* Available: http://www.compadre.org/ Repository/document/ServeFile.cfm?ID=9845&DocID=1498 [September 2013].

Nehm, R., and Härtig, H. (2011). Human vs. computer diagnosis of students› natural selection knowledge: Testing the efficacy of text analytic software. *Journal of Science Education and Technology, 21*(1), 56-73.

Newmann, F.M., Lopez, G., and Bryk, A. (1998). *The quality of intellectual work in Chicago schools: A baseline report.* Prepared for the Chicago Annenberg Research Project. Chicago, IL: Consortium on Chicago School Research. Available: https://ccsr. uchicago.edu/sites/default/files/publications/p0f04.pdf [April 2014].

Newmann, F. M., and Associates. (1996). *Authentic achievement: Restructuring schools for intellectual quality.* San Francisco, CA: Jossey-Bass.

NGSS Lead States. 2013. *Next generation science standards: For states, by states.* Washington, DC: Achieve, Inc. on behalf of the twenty-six states and partners that collaborated on the NGSS.

OECD. (2011). *Quality time for students: Learning in and out of school.* Paris, France: Author.

Pashler, H., Bain, P., Bottge, B., Graesser, A., Koedinger, K., McDaniel, M., and Metcalfe, J. (2007). *Organizing Instruction and Study to Improve Student Learning (NCER 2007-2004).* Washington, DC: National Center for Education Research, Institute of Education Sciences, U.S. Department of Education.

Pecheone, R., Kahl, S., Hamma, J., and Jaquith, A. (2010). *Through a looking glass: Lessons learned and future directions for performance assessment.* Stanford, CA: Stanford University, Stanford Center for Opportunity Policy in Education.

Pellegrino, J.W. (2013). Proficiency in science: Assessment challenges and opportunities. *Science, 340*(6130), 320-323.

Pellegrino, J.W., and Quellmalz, E.S. (2010-2011). Perspectives on the integration of technology and assessment. *Journal of Research on Technology in Education, 43*(3), 119-134.

Pellegrino, J.W., DiBello, L.V., and Brophy, S.P. (2014). The science and design of assessment in engineering education. In A. Johri and B.M. Olds (Eds.), *Cambridge handbook of engineering education research* (Ch. 29). Cambridge, UK: Cambridge University Press.

Penuel, W.R., Moorthy, S., DeBarger, A.H., Beauvineau, Y., and Allison, K. (2012). *Tools for orchestrating productive talk in science classrooms.* Workshop presented at the International Conference for Learning Sciences, Sydney, Australia.

Perie, M., Marion, S., Gong, B., and Wurtzel, J. (2007). *The role of interim assessment in a comprehensive assessment system.* Washington, DC: The Aspen Institute.

Peters, V., Dewey, T., Kwok, A., Hammond, G.S., and Songer, N.B. (2012). Predicting the impacts of climate change on ecosystems: A high school curricular module. *The Earth Scientist, 28*(3), 33-37.

Queensland Studies Authority. (2010a). *Moderation handbook for authority subjects.* The State of Queensland: Author.

Queensland Studies Authority. (2010b). *School-based assessment: The Queensland system.* The State of Queenlsand: Author.

Quellmalz, E.S., Timms, M.J., and Buckley, B.C. (2009). *Using science simulations to support powerful formative assessments of complex science learning.* Washington, DC: WestEd.

Quellmalz, E.S., Timms, M.J., Silberglitt, M.D., and Buckley, B.C. (2012). Science assessments for all: Integrating science simulations into balanced state science assessment systems. *Journal of Research in Science Teaching, 49*(3), 363-393.

Reiser, B.J. (2004). Scaffolding complex learning: The mechanisms of structuring and problematizing student work. *Journal of the Learning Sciences: 13*(3), 273-304.

Reiser, B.J. et al. (2013). Unpublished data from IQWST 6th grade classroom, collected by Northwestern University Science Practices project, PI Brian Reiser.

Rennie Center for Education Research and Policy. (2008). *Opportunity to learn audit: High school science.* Cambridge, MA: Author. Available: http://renniecenter.issuelab.org/resource/opportunity_to_learn_audit_high_school_science [December 2013].

Rosebery, A., and Warren, B. (Eds.). (2008). *Teaching science to English language learners: Building on students' strengths.* Arlington, VA: National Science Teachers Association.

Rosebery, A., Ogonowski, M., DiSchino, M. and Warren, B. (2010). The coat traps all the heat in your body: Heterogeneity as fundamental to learning. *Journal of the Learning Sciences 19*(3), 322-357.

Ruiz-Primo, M.A., Shavelson, R.J., Hamilton, L.S., and Klein, S. (2002). On the evaluation of systemic science education reform: Searching for instructional sensitivity. *Journal of Research in Science Teaching, 39*(5), 369-393.

Rutstein, D., and Haertel, G. (2012). *Scenario-based, technology-enhanced, large-scale science assessment task.* Paper prepared for the Technology-Enhanced Assessment Symposium, Educational Testing Service, SRI International, April 9.

Scalise, K. (2009). *Computer-based assessment: "Intermediate constraint" questions and tasks for technology platforms.* Available: http://pages.uoregon.edu/kscalise/taxonomy/taxonomy.html [October 2013].

Scalise, K. (2011). *Intermediate constraint taxonomy and automated scoring approaches.* Session presented for Colloquium on Machine Scoring: Specification of Domains, Tasks/Tests, and Scoring Models, the Center for Assessment, May 25-26, Boulder, CO.

Scalise, K., and Gifford, B.R. (2006). Computer-based assessment in e-learning: A framework for constructing "intermediate constraint" questions and tasks for technology platforms. *Journal of Teaching, Learning and Assessment, 4*(6).

Schmeiser, C.B., and Welch, C.J. (2006). Test development. In R.L. Brennan (Ed.), *Educational Measurement* (fourth ed.). Westport, CT: American Council on Education and Praeger.

Schmidt, W. et al. (1999). *Facing the consequences: Using TIMSS for a closer look at United States mathematics and science education.* Hingham, MA: Kluwer Academic.

Schwartz, R., Ayers, E., and Wilson, M. (2011). *Mapping a learning progression using unidimensional and multidimensional item response models.* Paper presented at the International Meeting of the Psychometric Society, Hong Kong.

Schwarz, C.V., Reiser, B.J., Davis, E.A., Kenyon, L., Acher, A., Fortus, D., Shwartz, Y., Hug, B., and Krajcik, J. (2009). Developing a learning progression for scientific modeling: Making scientific modeling accessible and meaningful for learners. *Journal of Research in Science Teaching, 46*(6), 632-654.

Shwartz, Y., Weizman, A., Fortus, D., Krajcik, J., and Reiser, B. (2008). The IQWST experience: Using coherence as a design principle for a middle school science curriculum. *The Elementary School Journal, 109*(2), 199-219.

Science Education for Public Understanding Program. (1995). *Issues, evidence and you: Teacher's guide.* Berkeley: University of California, Lawrence Hall of Science.

Shavelson, R.J., Baxter, G.P., and Gao, X. (1993). Sampling variability of performance assessments. *Journal of Educational Measurement, 30*(3), 215-232.

Shaw, J.M., Bunch, G.C., and Geaney, E.R. (2010). Analyzing language demands facing English learners on science performance assessments: The SALD framework. *Journal of Research in Science Teaching, 47*(8), 908-928

Smith, C., Wiser, M., Anderson, C., and Krajcik, J. (2006). Implications of research on children's learning for standards and assessment: A proposed learning progression for matter and atomic-molecular theory. *Measurement, 14*(1-2), 1-98.

Solano-Flores, G., and Li, M. (2009). Generalizability of cognitive interview-based measures across cultural groups. *Educational Measures: Issues and Practice, 28*(2), 9-18.

Solano-Flores, G., and Nelson-Barber, S. (2001). On the cultural validity of science assessments. *Journal of Research in Science Teaching, 38*(5), 553-573.

Songer, N.B. et al. (2013). Unpublished resource material from University of Michigan.

Songer, N.B., Kelcey, B., and Gotwals, A.W. (2009). How and when does complex reasoning occur? Empirically driven development of a learning progresion focused on complex reasoning about biodiversity. *Journal of Research in Science Teaching, 46*(6), 610-631.

SRI International. (2013). *Student assessment for everchanging earth unit.* Menlo Park, CA: Author.

Stecher, B.M., and Klein, S.P. (1997). The cost of science performance assessments in large-scale testing programs. *Educational Evaluation and Policy Analysis, 19*(1), 1-14.

Steinberg, L.S., Mislevy, R.J., Almond, R.G., Baird, A.B., Cahallan, C., DiBello, L.V., Senturk, D., Yan, D., Chernick, H., and Kindfield, A.C.H. (2003). *Introduction to the Biomass project: An illustration of evidence-centered assessment design and delivery capability.* CSE Report 609. Los Angeles: University of California Center for the Study of Evaluation.

Steinhauer, E., and Koster van Groos, J. (2013). *PISA 2015: Scientific literacy.* Available: http://www.k12center.org/rsc/pdf/s3_vangroos.pdf [December 2013].

Stiggins, R.J. (1987). The design and development of performance assessments. *Educational Measurement: Issues and Practice, 6,* 33-42.

Sudweeks, R.R., and Tolman, R.R. (1993). Empirical versus subjective procedures for identifying gender differences in science test items. *Journal of Research in Science Teaching, 30*(1), 3-19.

Thompson, S.J., Johnstone, C.J., and Thurlow, M.L. (2002). *Universal design applied to large scale assessments.* Synthesis Report 44. Minneapolis: University of Minnesota.

Topol, B., Olson, J., and Roeber, E. (2010). *The cost of new higher quality assessments: A comprehensive analysis of the potential costs for future state assessments.* Stanford, CA: Stanford University, Stanford Center for Opportunity Policy in Education.

Topol, B., Olson, J., Roeber, E., and Hennon, P. (2013). *Getting to higher-quality assessments: Evaluating costs, benefits, and investment strategies.* Stanford, CA: Stanford University, Stanford Center for Opportunity Policy in Education.

Tzou, C.T., and Bell, P. (2010). Micros and me: Leveraging home and community practices in formal science instruction. In K. Gomez, L. Lyons and J. Radinsky (Eds.), *Proceedings of the 9th International Conference of the Learning Sciences* (pp. 1135-1143). Chicago, IL: International Society of the Learning Sciences.

Tzou, C.T., Bricker, L.A., and Bell, P. (2007). *Micros and me: A fifth-grade science exploration into personally and culturally consequential microbiology.* Seattle: Everyday Science and Technology Group, University of Washington.

U.S. Census Bureau. (2012). *The 2012 statistical abstract.* Available: http://www.census.gov/compendia/statab/cats/education.html [September 2013].

Wainer, H. (2003). Reporting test results in education. In R. Fernández-Ballesteros (Ed.), *Encyclopedia of psychological assessment.* (pp. 818-826). London: Sage.

Warren, B., Ballenger, C., Ogonowski, M., Rosebery, A. and Hudicourt-Barnes, J. (2001). Rethinking diversity in learning science: The logic of everyday sensemaking. *Journal of Research in Science Teaching, 38,* 1-24.

Warren, B., Ogonowski, M. and Pothier, S. (2005). "Everyday" and "scientific": Re-thinking dichotomies in modes of thinking in science learning. In R. Nemirovsky, A. Rosebery, J. Solomon, and B. Warren (Eds.), *Everyday matters in science and mathematics: Studies of complex classroom events* (pp. 119-148). Mahwah, NJ: Lawrence Erlbaum Associates.

Wiggins, G., and McTight, J. (2005). *Understanding by design: Expanded 2nd edition.* New York: Pearson.

Williamson, D.M., Mislevy, R.J., and Bejar, I.I. (2006). *Automated scoring of complex tasks in computer-based testing.* Mahwah, NJ: Lawrence Erlbaum Associates.

Williamson, D.M., Xi, X., and Breyer, F.J. (2012). A framework for evaluation and use of automated scoring. *Educational Measurement: Issues and Practice, 31*(1), 2-13.

Wilson, M. (2005). *Constructing measures: An item response modeling approach.* Mahwah, NJ: Lawrence Erlbaum Associates.

Wilson, M. (2009). Measuring progressions: Assessment structures underlying a learning progression. *Journal for Research in Science Teaching, 46*(6), 716-730.

Wilson, M., and Draney, K. (2002). A technique for setting standards and maintaining them over time. In S. Nishisato, Y. Baba, H. Bozdogan, and K. Kanefugi (Eds.), *Measurement and multivariate analysis* (pp. 325-332). Proceedings of the International Conference on Measurement and Multivariate Analysis, Banff, Canada, May 12-14, 2000. Tokyo: Springer-Verlag.

Wilson, M., and Scalise, K. (2012). *Measuring collaborative digital literacy.* Paper presented at the Invitational Research Symposium on Technology Enhanced Assessments, May 7-8, Washington, DC. Available: http://www.k12center.org/rsc/pdf/session5-wilson-paper-tea2012.pdf [April 2014].

Wilson, M. and Sloane, K. (2000). From principles to practice: An embedded assessment system. *Applied Measurement in Education, 13*(2), 181-208.

Wilson, M, and Wang, W. (1995). Complex composites: Issues that arise in combining different modes of assessment. *Applied Psychological Measurement. 19*(1), 51-72.

Wilson, M. et al. (2013). Unpublished data from the BEAR Center at the University of California, Berkeley.

Wood, W.B. (2009). Innovations in teaching undergraduate biology and why we need them. *Annual Review of Cell and Developmental Biology, 25,* 93-112.

Yen, W.M., and Ferrara, S. (1997). The Maryland school performance assessment program: Performance assessment with psychometric quality suitable for high-stakes usage. *Educational and Psychological Measurement, 57,* 60-84.

Young, B.J., and Lee, S.K. (2005). The effects of a kit-based science curriculum and intensive professional development on elementary student science achievement. *Journal of Science Education and Technology, 14*(5/6), 471-481.

WORKSHOP AGENDA

Workshop on Developing Assessments to Meet the Goals of the
2012 Framework for K-12 Science Education
September 13, 2012

National Academyof Sciences Building
2101 Constitution Ave., NW
Auditorium
Washington DC

AGENDA

8:30   Registration, check-in for workshop

9:00-9:15   Welcome, Introductions, Overview of the Agenda
              (9:00) Stuart Elliott, Director, Board on Testing and Assessment
              (9:05) Martin Storksdieck, Director, Board on Science Education
              (9:10) David Heil, Collaborative Mentor, CCSSO's State Collaborative
              on Assessment and Student Standards (SCASS) in Science

Part I: Problem Statement: Laying Out the Problem/Challenges

This session will review the Framework and what it calls for and discuss the challenges that it poses for assessment.

Moderator: Mark Wilson, University of California at Berkeley, Committee Cochair

9:15-10:15   **What is the vision of learning and instruction laid out in the Framework? What are the implications for assessment?**
(9:15) Helen Quinn, Stanford University, Committee Member
(9:35) Jim Pellegrino, University of Illinois at Chicago, Committee Cochair

Reactions and Questions
(9:55) James Woodland, Nebraska Department of Education
(10:00) Robin Anglin, West Virginia Department of Education
(10:05) Audience Q and A

10:15-10:30   Break

Part II: Exploring Alternatives: Strategies for Assessing Learning as Envisioned in the Framework

Assessing the proficiencies depicted in the Framework will require changes to the status quo. Innovative assessment formats and technology enhancements may offer the means for assessing some of the skills and performances on large-scale, external tests. Some of the skills and performances may not be well suited to large-scale, external testing formats, but other ways of measuring them may produce results that can be utilized in new ways. This session will focus in detail on some of the alternatives.

10:30-12:00   Large-Scale Assessments
In this session a series of panelists will discuss examples of large-scale assessments that assess science practices in conjunction with core ideas and crosscutting concepts, similar to those depicted in the Framework. Focus will be on how these strategies can be used to measure learning as envisioned in the Framework.

Moderators:
Catherine Welch, University of Iowa, Committee Member
Kathleen Scalise, University of Oregon, Committee Member

Presenters will address the following questions:

1. How are content knowledge, crosscutting concepts, and science practices assessed in the program? If possible, please provide one or more sample tasks and discuss the content and practices that are assessed.
2. How is the assessment administered? How long does it take and what materials and/or technologies are needed?
3. How are the tasks scored and how are scores reported? Are scores reported separately for content knowledge, crosscutting concepts, and practices or is a composite score created?
4. What steps, if any, are taken to ensure that scores are comparable from one administration to the next?
5. What was involved in developing the assessment tasks/items? What challenges were encountered and how were they handled? Please discuss any practical, cost, or feasibility issues that arose and how they were addressed.

(10:30) **NAEP 2009 Science Assessment: Hands-On and Interactive Computer Tasks**
Alan Friedman, National Assessment Governing Board
Peggy Carr, National Center for Education Statistics
(10:50) **College Board's Advanced Placement Tests in Biology**
Rosemary Reshetar, College Board
(11:10) **SimScientists**
Edys Quellmalz, WestEd

Reactions and Questions
(11:30) Moderators' follow-up questions to panelists
(11:40) Yvette McCulley, Iowa Department of Education
(11:50) Audience Q and A

12:00-12:45   Lunch in Great Hall

12:45-2:30   **Assessments Embedded in Curricular Units**
The Framework calls for an approach to instruction and assessment that utilizes learning progressions and associated curricular units. What assessment strategies can be used to measure students' achievement in relation to a learning progression? What types of activities/tasks allow us to make inferences about where a student is on the progression? This session will feature examples of work to develop assessments of learning progressions in conjunction with curricular units.

Moderator: Mark Wilson
(12:45) Introductory Remarks by the Moderator

Assessing Science Knowledge That Inextricably Links Core Disciplinary Ideas and Practices
(1:00) Joe Krajcik, Michigan State University
(1:15) Nancy Butler Songer, University of Michigan, Committee Member
(1:30) Brian Reiser, Northwestern University, Committee Member
(1:45) Rich Lehrer, Vanderbilt University, Committee Member

Reactions and Questions
(2:00) Roberta Tanner, Loveland High School, Committee Member
(2:10) Beverly Vance, North Carolina Department of Public Instruction
(2:20) Audience Q and A

2:30-3:15   **Measurement Challenges**
This session will consider the featured sample assessments—both large-scale and curriculum-embedded—and discuss the measurement challenges associated with these approaches. The session will focus on issues such as: (1) to what extent do these approaches offer viable alternatives for assessing science learning consistent with the Framework; (2) to what extent are these approaches likely to yield scores that support the desired inferences and policy purposes; (3) what practical, technical, and psychometric challenges might arise with these approaches?

Moderator: Mark Wilson

 (2:30) Ed Haertel, Stanford University, Committee Member

Reactions and Questions
(2:50) Anita Bernhardt, Maine Department of Education
(2:57) Jeff Greig, Connecticut State Department of Education
(3:05) Audience Q and A

3:15-3:30   Break

Part III: Developing Systems of Assessments

This session will address different strategies for gathering assessment information—some based on summative assessment, some based on end-of-course assessments, and some based on collections of classroom work—and consider how to integrate/combine the information. The session will discuss models used in other countries and settings that provide ways to integrate a broad range of assessment information.

3:30-4:30   **Moderator: Jerome Shaw, University of California, Santa Cruz, Committee Member**

Presenters:
(3:30) Joan Herman, CRESST, Committee Member
(3:45) Knut Neumann, University of Kiel, Committee Member

Reactions and Questions:
(4:00) Susan Codere Kelly, Michigan Department of Education
(4:10) Melinda Curless, Kentucky Department of Education
(4:20) Audience Q and A

Part IV: Synthesis

4:30-5:45   **Moderators: Jim Pellegrino, Mark Wilson**

Panel
(4:30) Peter McLaren, Rhode Island Department of Elementary and Secondary Education, Committee Member

(4:40) Richard Amasino, University of Wisconsin–Madison, Committee Member

(4:50) Shelley Lee, Wisconsin Department of Public Instruction

(5:00) Matt Krehbiel, Kansas State Department of Education

(5:10) Comments from the Moderators

(5:20) Audience Q and A

Questions for Discussion

- What are the main takeaway points from the workshop discussions?
- Considering the sample assessments discussed during the workshop, which approaches to assessment seem most promising and consistent with the goals of the Framework? What challenges do they help solve? What challenges would still need to be solved?
- What additional issues should the committee explore?

5:45   **Adjourn**

B

BIOGRAPHICAL SKETCHES OF COMMITTEE MEMBERS AND STAFF

COMMITTEE

James W. Pellegrino (*Cochair*) is liberal arts and sciences distinguished professor and distinguished professor of education at the University of Illinois at Chicago (UIC). He also serves as codirector of UIC's Interdisciplinary Learning Sciences Research Institute. Dr. Pellegrino's research and development interests focus on children's and adult's thinking and learning and the implications of cognitive research and theory for assessment and instructional practice. Much of his current work is focused on analyses of complex learning and instructional environments, including those incorporating powerful information technology tools, with the goal of better understanding the nature of student learning and the conditions that enhance deep understanding. A special concern of his research is the incorporation of effective formative assessment practices, assisted by technology, to maximize student learning and understanding. Increasingly, his research and writing has focused on the role of cognitive theory and technology in educational reform and translating results from the educational and psychological research arenas into implications for practitioners and policy makers. Dr. Pellegrino has served on numerous National Research Council (NRC) boards and committees, including the Board on Testing and Assessment. He cochaired the NRC committee that authored the report *Knowing What Students Know: The Science and Design of Educational Assessment*. Most recently, he served as a member of the Committee on Conceptual Framework for New Science Education Standards, as well as the Committee on Test Design for K-12 Science Achievement, and the Committee on Science Learning: Games, Simulations and Education. He is a fellow of the American Educational Research Association, and a lifetime national associate of the National Academy of Sciences, and in 2007 he was elected to lifetime membership in the National Academy of Education. Dr. Pellegrino earned his B.A. in psychology from Colgate University and both his M.A. and Ph.D. from the University of Colorado.

Mark R. Wilson (*Cochair*) is professor of policy, organization, measurement, and evaluation in the Graduate School of Education at University of California, Berkeley. He is also the founder and director of the Berkeley Evaluation and Assessment Research Center. His main research areas are educational measurement, psychometrics, assessment design, and applied statistics. His current research is focused on (a) developing assessments and psychometric models for learning progressions, especially assessments that are technologically enhanced, and (b) rethinking the philosphical foundations of measurement in the social sciences. He is founding editor of the journal, *Measurement: Interdisciplinary Research and Perspectives*, and has recently served as president of the Psychometric Society. Dr. Wilson has extensive experience with National Research Council projects. He served on the Committee on the Foundations of Assessment; the Committee on Development Outcomes and Assessment for Young Children; the Committee on Value-Added Methodology for Instructional Improvement, Program Evaluation, and Accountability; and the Committee on Best Practices for State Assessment Systems: Improving Assessment While Revisiting Standards. He chaired the Committee on Test Design for K-12 Science Achievement and currently serves on the Board on Testing and Assessment. He is a fellow of the American Psychological Association, and the American Educational Research Association, is a national associate of the National Academy of Sciences, and in 2011 was elected to membership in the National Academy of Education. Dr. Wilson has a Ph.D. in measurement and educational statistics from the University of Chicago.

Richard M. Amasino is Howard Hughes Medical Institute (HHMI) professor with the Department of Biochemistry at the University of Wisconsin–Madison. His research addresses the mystery of how a plant knows that it has been through a complete winter and that it is now safe to flower in response to the lengthening days of spring. Now, as an HHMI professor, the plant biologist plans to use plant genetics to involve undergraduates in original experiments and to develop appealing, accessible genetics-based teaching units for K-12 science. He has received numerous awards in biological science and was elected as a National Academy of Sciences member in 2006. With the National Research Council, he is currently chair of Section 62: Plant, Soil, and Microbial Sciences, as well as a section representative for the 2012 NAS Class VI Membership Committee. Dr. Amasino received his B.S. in biology from Pennsylvania State University and his M.S. and Ph.D. in biology/biochemistry from Indiana University.

Edward H. Haertel is Jacks Family professor of education (emeritus) at the Graduate School of Education at Stanford University. His research centers on policy uses of achievement test data, including examination of value-added models for teacher evaluation from a psychometric perspective. Dr. Haertel has been closely involved in the creation and maintenance of California's school accountability system both before and after passage of the No Child Left Behind Act. In addition to technical issues in designing accountability systems and quantifying their precision, his work is concerned with validity arguments for high-stakes testing, the logic and implementation of standard-setting methods, and comparisons of trends on different tests and in different reporting metrics. He has served as president of the National Council on Measurement in Education and as a member of the National Assessment Governing Board. He is currently serving as chair of the Board on Testing and Assessment and previously was a member of the Committee on Review of Alternative Data Sources for the Limited-English Proficiency Allocation Formula under Title III, Part A, Elementary and Secondary Education Act. He has served on numerous state and national advisory committees related to educational testing, assessment, and evaluation, including the joint committee responsible for the 1999 revision of the *Standards for Educational and Psychological Testing*. He currently serves on the technical advisory committee for the Smarter Balanced Assessment Consortium, funded by the Race to the Top initiative. He has been a fellow at the Center for Advanced Study in the Behavioral Sciences, is a fellow of the American Psychological Association, and is a member of the National Academy of Education. Dr. Haertel holds a Ph.D. in measurement, evaluation, and statistical analysis from the University of Chicago.

Joan Herman is senior research scientist of the National Center for Research on Evaluation, Standards, and Student Testing at the University of California, Los Angeles. Her research has explored the effects of testing on schools and the design of assessment systems to support school planning and instructional improvement. Her recent work has focused on the validity and utility of teachers' formative assessment practices in mathematics and science. She also has wide experience as an evaluator of school reform and is noted in bridging research and practice. She is past president of the California Educational Research Association; has held a variety of leadership positions in the American Educational Research Association and Knowledge Alliance; is a member of the joint committee for the Revision of the Standards for Educational and Psychological Measurement; cochair of the Board of Education for Para Los Niños; and is current editor of *Educational Assessment*. Dr. Herman currently serves on the technical advisory committee for the Smarter Balanced Assessment Consortium, funded by the U.S. Department of Education's Race to the Top initiative. She has extensive experience serving on National Research Council projects. She is currently a member of the Board on Testing and Assessment (BOTA). She served as a member of the Committee on Test Design for K-12 Science Achievement, the Roundtable on Education Systems and Accountability, and the Committee on Best Practices for State Assessment Systems, and, most recently, chaired the BOTA workshop on 21st Century Skills. Dr. Herman received her doctorate of education in learning and instruction from the University of California, Los Angeles.

Richard Lehrer is professor of science education in the Department of Teaching and Learning at Peabody College of Vanderbilt University. Previously, he has taught in a number of different settings from high school science to the university level. He was also associate director of the National Center for Improving Student Learning and Achievement in Mathematics and Science as well as associate director of the National Center for Research in Mathematical Sciences Education. His research focuses on children's mathematical and scientific reasoning in the context of schooling, with a special emphasis on tools and notations for developing thought. Dr. Lehrer has been on a number of National Research Council (NRC) committees covering K-12 science education and achievement, including the Committee on Test Design for K-12 Science Achievement. He is currently a member of the NRC study Toward Integrating STEM Education: Developing a Research Agenda. Dr. Lehrer received his B.S. in biology and chemistry from Rensselaer Polytechnic Institute, and his M.S. and Ph.D. in educational psychology and statistics from the University of New York at Albany.

Scott F. Marion is the vice president of the National Center for the Improvement of Educational Assessment, Inc., where his current projects include developing and implementing reform-based educator evaluation systems, designing validity evaluations for state assessment and accountability systems, including teacher evaluation systems, and designing and implementing high-quality, locally designed performance-based assessments. He also is a leader in designing approaches for documenting evidence of student learning for teachers in nontested subjects and grades in valid and educationally supportive ways. He coordinates and/or serves on multiple state technical advisory committees, is the coordinator of the Partnership for Assessment of Readiness for College and Careers assessment consortium technical advisory committee, and was a former member of the U.S. Department of Education's National Technical Advisory Committee. Dr. Marion previously served on the National Research Council's Committee on Value-Added Methodology for Instructional Improvement, Program Evaluation, and Accountability, and the Committee on Best Practices for State Assessment Systems. Prior to joining the Center for Assessment in early 2003, he was most recently the director of assessment and accountability for the Wyoming Department of Education. Dr. Marion regularly presents the results of his work at several national conferences (American Educational Research Association, National Council on Measurement in Education, and Council of Chief State School Officers) and has published dozens of articles in peer-reviewed journals and edited volumes. He is a member of his local school board in Rye, New Hampshire. A former field biologist and high school science teacher, Dr. Marion earned a B.S. in biology from the State University of New York and an M.S. in science education from the University of Maine. Dr. Marion received his Ph.D. in measurement and evaluation from the University of Colorado, Boulder.

Peter McLaren is a science and technology specialist at the Rhode Island Department of Education, where he has participated in a number of activities related to the Next Generation Science Standards (NGSS). He also codirects the administration of the New England Common Assessment Program science assessments and cofacilitates Rhode Island's NGSS Strategic Leadership Council. Mr. McLaren is also past president of the Council of State Science Supervisors and currently serves as a member of the NGSS Writing Team for Achieve. Previously, he was a science teacher for 13 years at both the high school and middle school levels. As an educator, McLaren was recognized with the Milken Family Foundation National Educator Award (2001) and as the Rhode Island Science Teacher of the Year (1995) by the Network of Educators of Science and Technology. McLaren has a B.S. in secondary education, and an M.A. in science education, both from the University of Rhode Island.

Knut Neumann is director of the Department of Physics Education at the Leibniz Institute for Science and Mathematics Education and professor of physics education at the University of Kiel, Germany. Previously, he worked in the Research

Group and Graduate School Teaching and Learning of Science at the University of Duisburg-Essen, where he was a member of the group of researchers who developed what later became the assessment framework for benchmarking the National Education Standards for the science subjects in Germany. During his career, Dr. Neumann developed a special interest in assessment. He is currently involved with several projects focusing on the assessment of students understanding of core physics concepts (e.g., energy and matter) and practices (e.g., performing experiments), teachers' professional knowledge (e.g., content knowledge and pedagogical content knowledge), and vocational knowledge for physics-related vocations. The ultimate goals of his activities are the development and empirical validation of learning progressions for K-12 physics and, based on these learning progressions, the improvement of instructional quality in physics through teacher professionalization. Dr. Neumann studied mathematics and physics for the teaching profession at the University of Düsseldorf and holds a Ph.D. from the University of Education at Heidelberg.

William Penuel is professor in educational psychology and the learning sciences at the University of Colorado at Boulder. Prior to this he was a director of evaluation research with SRI International's Center for Technology in Learning. Dr. Penuel's research focuses on teacher learning and organizational processes that shape the implementation of educational policies, school curricula, and afterschool programs. One strand of his research focuses on designs for teacher professional development in Earth science education. A second strand examines the role of research-practice partnerships in designing supports for teacher learning in school districts. A third strand focuses on the development of science-linked interests and identities among children and youth. He is currently on the editorial board for *Teachers College Record, American Journal of Evaluation,* and *Cognition and Instruction.* Dr. Penuel received his Ph.D. in developmental psychology from Clark University.

Helen R. Quinn is professor emerita of particle physics and astrophysics at SLAC National Accelerator Laboratory. She has taught physics at both Harvard and Stanford. Dr. Quinn is an internationally recognized theoretical physicist who holds the Dirac Medal (from the International Center for Theoretical Physics, Italy), the Klein Medal (from the Swedish National Academy of Sciences and Stockholm University) and the Sakurai Prize (from the American Physical Society). She is a member of the American Academy of Arts and Sciences, the National Academy of Sciences, and the American Philosophical Society. She is a fellow and former president of the American Physical Society. She is originally from Australia and is an honorary officer of the Order of Australia. Dr. Quinn is chair of the National Academy of Sciences Board on Science Education (BOSE). She served as a member of the BOSE study that developed the report *Taking Science to School* and led the committee for *A Framework for K-12 Science Education*, which are the basis of the Next Generation Science Standards that have now been adopted by multiple states in the United States. Dr. Quinn received her Ph.D in physics at Stanford in 1967.

Brian J. Reiser is professor of learning sciences in the School of Education and Social Policy at Northwestern University. His research examines how to make scientific practices such as argumentation, explanation, and modeling meaningful and effective for classroom teachers and students. This design research investigates the cognitive and social interaction elements of learning environments supporting scientific practices, and design principles for technology-infused curricula that embed science learning in investigations of contextualized data-rich problems. Dr. Reiser is also on the leadership team for IQWST (Investigating and Questioning our World through Science and Technology), a collaboration with the University of Michigan, developing a middle school project-based science curriculum. He was a founding member of the first graduate program in learning sciences, created at Northwestern, and chaired the program from 1993, shortly after its inception, until 2001. He was coprincipal investigator in the National Science Foundation Center for Curriculum Materials in Science, exploring the design and enactment of science curriculum materials. His National Research Council work includes the recent Committee on a Conceptual Framework for New Science Education Standards and the committee that authored *Taking Science to School*. Dr. Reiser received his Ph.D. in cognitive science from Yale University.

Kathleen Scalise is an associate professor at the University of Oregon in the Department of Educational Methodology, Policy and Leadership. Her main research areas are technology-enhanced assessments in science and mathematics education, item-response models with innovative item types, dynamically delivered content in e-learning, computer adaptive testing, and applications to equity studies. She recently served as a core member of the methodological group for the Assessment and Teaching of 21st Century Skills project created by Cisco, Intel, and Microsoft; for the Oregon state task force, writing legislation for virtual public schools; as codirector of the University of California, Berkeley, Evaluation and Assessment Research Center, and for the U.S. Department of Education on the

Race to the Top Assessment Program competition. She has been a visiting scholar in the Department of Chemistry at the University of California, Berkeley, and a visiting scientist in the Department of Neuroscience at Columbia University in 2012-2013. She currently is on the expert's group for the *Program for International Student Assessment* 2015, which has major domain focus in science education, as well as assessments in collaborative problem solving for the 2015 assessment cycle. She also served with the Curriculum Frameworks and Instructional Resources Division of the California Department of Education for development of the state science framework. Dr. Scalise holds teaching credentials for K-12 physical and life sciences and has experience in middle and secondary science instruction as well as at the postsecondary and graduate education levels in measurement, statistics, instructional technology, and analysis of teaching and learning. Dr. Scalise received her Ph.D. in quantitative measurement at the University of California, Berkeley, in 2004.

Jerome M. Shaw is an associate professor of science education at the University of California, Santa Cruz. He has more than 30 years of experience in education with a focus on understanding and improving science teaching and learning for culturally and linguistically diverse students. As a classroom teacher in California public schools, Dr. Shaw taught science at the elementary and secondary levels in mainstream, bilingual (Spanish-English), and structured English immersion classrooms. His research examines science teaching and learning for culturally and linguistically diverse students with an explicit focus on the relationship of assessment to this larger process. Conceptually, his research agenda explores the overlap among science teaching and learning, assessment of student learning, and equity and diversity issues in education. The unifying theme across these intersections is a focus on English-language learners. Operationally, his research program is organized along four strands: (1) clarifying the nature of the achievement gap, (2) identifying fairness issues posed by assessment practices, (3) developing new performance assessments, and (4) enhancing the ability of teachers to provide effective instruction and assessment. These strands, though distinct, are interrelated and complementary. He holds lifetime California teaching credentials for high school biology, Spanish, and social studies, as well as multiple elementary subjects coupled with a certificate of bilingual-bicultural competency. Dr. Shaw received a B.A. in Spanish, an M.A. in education, and a doctorate in science education, all from Stanford University.

Nancy Butler Songer is a professor at the University of Michigan, Ann Arbor, and the director of the Center for Essential Science, an interdisciplinary research center at the University of Michigan. Songer's research examines science learning through the creation and evaluation of curricular and technological resources focused on the study of some of the most pressing environmental issues of our time, such as climate change and biodiversity, through the science and engineering practices of data analysis, argumentation, and the use of models to make predictions.Recognition includes fellow of the American Association for the Advancement of Science and selection by the U.S. Secretary of Education for the Promising Educational Technology Award. In 1995, she received a National Science Foundation Presidential Faculty Fellowship from President Clinton, the first science educator to receive this recognition. Prior to coming to Michigan in 1996, Dr. Songer earned an M.S. in developmental biology from Tufts University and a Ph.D. in science education from the University of California, Berkeley.

Roberta Tanner is a physics teacher at Loveland high school in Colorado. She has a keen interest in science and engineering education and a fascination with understanding how people learn. She taught physics, math, engineering, and other science courses for 21 years at a high school in the Thompson School District in Loveland, Colorado. Wanting to spur her students to higher levels of achievement, she brought advanced placement physics and integrated physics/trigonometry to the district and taught those for 15 years. She also designed and taught Microcomputer Projects—an award winning project-oriented microchip and electrical engineering course. In addition, she was privileged to work for a year as Teacher in Residence with the Physics Education Research group at the University of Colorado, Boulder. There she learned a great deal about how students learn. She also taught introductory physics at the University of Colorado. Tanner was honored with the International Intel Excellence in Teaching Award in 2004 and the Amgen Award for Science Teaching Excellence in 2011. She served 5 years on the Teacher Advisory Council, an advisory board to the National Academy of Sciences. She also served on a committee of the National Academy of Engineering, investigating the advisability of National K-12 Engineering Standards. Tanner completed her undergraduate work in physics and mechanical engineering at Kalamazoo College and Michigan State University. She earned her teaching certificate and a master's degree in education at the University of Colorado, Boulder.

Catherine J. Welch is professor with the Department of Psychological and Quantitative Foundations and Educational Measurement and Statistics Program at the University of Iowa. In addition to teaching courses in educational measurement and conducting measurement-related research, Dr. Welch codirects the Iowa Testing Programs. Prior to joining the faculty at the University of Iowa, she served as an assistant vice president with ACT, where she worked on a variety of assessment programs for over 22 years, predominantly with ACT's Performance Assessment Center. At ACT, Welch worked with state and national education officials and measurement experts on a broad range of testing issues and became widely recognized as an authority on large-scale assessments. Her research interests include educational assessment, college readiness, validity evaluation, and educational measurement and statistics. Welch has served on the board of directors for the National Council on Measurement in Education, and she recently received the distinguished research award through the Iowa Educational Research and Evaluation Association. Dr. Welch received her M.A. and Ph.D. in educational measurement and statistics from the University of Iowa.

STAFF

Alexandra Beatty is a senior program officer for the Board on Testing and Assessment. Since 1996 she has contributed to many projects, including an evaluation of the District of Columbia Public Schools; studies of teacher preparation, National Board certification for teachers, and state-level science assessment; and the Committee on Education Excellence and Testing Equity. She has also worked as an independent education writer and researcher. Prior to joining the National Research Council staff, she worked on the National Assessment of Educational Progress and College Board programs at the Educational Testing Service. She has a B.A. in philosophy from Williams College and an M.A. in history from Bryn Mawr College.

Stuart Elliott is director of the Board on Testing and Assessment at the National Research Council where he has worked on a variety of projects related to assessment, accountability, teacher qualifications, and information technology. Previously, he worked as an economic consultant for several private-sector consulting firms. He was also a research fellow in cognitive psychology and economics at Carnegie Mellon University, and a visiting scholar at the Russell Sage Foundation. He has a Ph.D. in economics from the Massachusetts Institute of Technology.

Judith A. Koenig is a senior program officer with the National Research Council's Board on Testing and Assessment where, since 1999, she has directed measurement-related studies designed to inform education policy. Her work has included studies on the National Assessment of Educational Progress; teacher licensure and advanced-level certification; inclusion of special-needs students and English-language learners in assessment programs; developing assessments for state and federal accountability programs in K-12 and adult education; setting standards for the National Assessment of Adult Literacy; assessing 21st century skills; and using value-added methods for evaluating schools and teachers. Previously, from 1984 to 1999, she worked at the Association of American Medical Colleges on the Medical College Admission Test where she directed operational programs and led a comprehensive research program on the examination. Prior to that, she worked for 10 years as a special education teacher and diagnostician. She received a B.A. in special education from Michigan State University, an M.A. in psychology from George Mason University, and a Ph.D. in educational measurement, statistics, and evaluation from the University of Maryland.

Heidi Schweingruber is the deputy director of the Board on Science Education (BOSE) at the National Research Council (NRC). In this role, she oversees many of the projects in the BOSE portfolio. She also collaborates with the director and board to develop new projects. She codirected the study that resulted in the report *A Framework for K-12 Science Education: Practices, Crosscutting Concepts, and Core Ideas* (2012), which is the first step in revising national standards for K-12 science education. She served as study director for a review of NASA's pre-college education programs completed in 2008 and codirected the study that produced the 2007 report *Taking Science to School: Learning and Teaching Science in Grades K-8*. She served as an editor on the NRC report *Mathematics Learning in Early Childhood: Paths to Excellence and Equity* (2009). She coauthored two award-winning books for practitioners that translate findings of NRC reports for a broader audience: *Ready, Set, Science! Putting Research to Work in K-8 Science Classrooms* (2008) and *Surrounded by Science* (2010). Prior to joining the NRC, she worked as a senior research associate at the Institute of Education Sciences in the U.S. Department of Education where she administered the preschool curriculum evaluation program and a grant program in mathematics education. Previously, she was the director of research for the Rice University School Mathematics Project, an outreach program in K-12 mathematics education, and taught in the psychology and education departments at Rice University. She holds a Ph.D. in psychology (developmental) and anthropology, and a certificate in culture and cognition from the University of Michigan.

Martin Storksdieck is the director of the Board on Science Education at the National Research Council (NRC) and the NRC's Roundtable on Climate Change Education. He oversees studies that address a wide range of issues related to science education and science learning, and provides evidence-based advice to decision makers in policy, academia, and educational practice. His prior research focused on what and how we learn when we do so voluntarily, and how learning is connected to our behaviors, identities, and beliefs. This includes the role of personal perspectives in science learning, particularly related to controversial topics such as climate change or evolution, and how connections between school-based and out-of-school learning can create and sustain lifelong interest in science and learning. Storksdieck's research also focused on the role of science-based professionals and science hobbyists in communicating their passions to a broader public. Before joining the NRC, he served as director of project development and senior researcher at the nonprofit Institute for Learning Innovation. In the 1990s, he was a science educator with a planetarium in Germany, where he developed shows and programs on global climate change; served as editor, host, and producer for a weekly environmental news broadcast; and worked as an environmental consultant specializing in local environmental management systems. He holds an M.S. in biology from the Albert-Ludwigs University in Freiburg, Germany; an M.P.A. from Harvard University's Kennedy School of Government; and a Ph.D. in education from Leuphana University in Lüneburg, Germany.